KU-714-175

Eukaryotic Transcription Factors

Second Edition

David S. Latchman

Professor of Molecular Pathology, University College London Medical School, London

ACADEMIC PRESS

Harcourt Brace & Company, Publishers

London   San Diego   New York
Boston   Sydney   Tokyo   Toronto

ACADEMIC PRESS LIMITED
24–28 Oval Road
LONDON NW1 7DX

U.S. Edition Published by
ACADEMIC PRESS INC.
San Diego, CA 92101

This book is printed on acid-free paper

Copyright © 1995 ACADEMIC PRESS LIMITED

All rights reserved

No part of this book may be reproduced or transmitted in any form or by
any means, electronic or mechanical including photocopying, recording,
or any information storage and retrieval system without permission in
writing from the publisher.

A catalogue record for this book is available from the British Library

ISBN 0–12–437172–8
ISBN 0–12–437173–6 (paperback)

Typeset by J&L Composition Ltd, Filey, North Yorkshire
Printed and bound in Great Britain by The University Press, Cambridge

Contents

To Maurice and Vivienne
In admiration

Preface to the second edition

In the four years since the first edition of this work was published, the explosion of information about transcription factors has continued. The genes encoding many more transcription factors have been cloned and this information used to analyse their structure and function, culminating in many cases with the use of inactivating mutations to prepare so-called 'knock-out' mice, thereby testing directly the role of these factors in development. None the less, the examples used in the first part of this book to illustrate the role of transcription factors in processes as diverse as inducible gene expression and development still remain amongst the best understood. The discussions of these factors has therefore been considerably updated to reflect the progress made in the past few years. In addition new sections have been added on topics such as TPB, the *myc* oncogene and anti-oncogenes where the degree of additional information now warrants a separate section.

Even greater changes have been necessary in the second part of the book, which deals with the mechanisms by which transcription factors act. Thus, for example, the sections in Chapter 9 on the mechanisms of transcriptional activation and on transcriptional repression have been completely rewritten. In addition the increasing emphasis on transcriptional repression discussed in Chapter 9 has led to a change in the title of Chapter 10 to 'What regulates the regulators?' (from 'What activates the activators?'). Moreover, this chapter now includes a much more extensive section on the interaction between different factors, which is another major theme to have emerged in the past few years. It is hoped that these changes will allow the new edition to build on the success of the first edition in providing an overview of these vital factors and the role they play in gene regulation.

Finally I would like to thank Jane Templeman who has prepared a large number of new illustrations to complement the excellent ones she provided for the first edition and Sarah Chinn for coping with the necessity of adding, deleting or amending large sections of the first edition. I am also grateful to Tessa Picknett and the staff at Academic Press for commissioning this new edition and their efficiency in producing it.

David S. Latchman

Preface to the first edition

In my previous book, *Gene Regulation: A Eukaryotic Perspective* (Unwin-Hyman Ltd, 1990) I described the mechanisms by which the expression of eukaryotic genes is regulated during processes as diverse as steroid treatment and embryonic development. Although some of this regulation occurs at the post-transcriptional level, it is clear that the process of gene transcription itself is the major point at which gene expression is regulated. In turn this has focused attention on the protein factors, known as transcription factors, which control both the basal processes of transcription and its regulation in response to specific stimuli or developmental processes. The characterization of many of these factors and in particular the cloning of the genes encoding them has resulted in the availability of a bewildering array of information on these factors, their mechanism of action and their relationship to each other. Despite its evident interest and importance, however, this information could be discussed only relatively briefly in *Gene Regulation*, whose primary purpose was to provide an overview of the process of gene regulation and the various mechanisms by which this is achieved.

It is the purpose of this book, therefore, to discuss in detail the available information on transcription factors, emphasizing common themes and mechanisms to which new information can be related as it becomes available. As such it is hoped the work will appeal to final-year undergraduates and postgraduate students entering the field as well as to those moving into the area from other scientific or clinical fields who wish to know how transcription factors may regulate the gene in which they are interested.

In order to provide a basis for the discussion of transcription factors, the first two chapters focus respectively on the DNA sequences with which the factors interact and on the experimental methods which are used to study these factors and obtain the information about them provided in subsequent chapters. The remainder of the work is divided into two distinct portions. Thus Chapters 3 to 7 focus on the role of transcription factors in particular processes. These include constitutive and inducible gene expression, cell type-specific and developmentally regulated gene expression and the role of transcription factors in cancer. Chapters 8 to 10 adopt a more mechanistic approach and consider the features of transcription factors which allow them to fulfil their function. These include the ability to bind to DNA and modulate transcription either positively or negatively as well as the ability to respond to specific stimuli and thereby activate gene expression in a regulated manner.

Although this dual approach to transcription factors from both a process-oriented and mechanistic point of view may lead to some duplication, it is the most efficient means of providing the necessary overview both of the nature of transcription factors and the manner in which they achieve their role of modulating gene expression in many diverse situations.

Finally I would like to thank Mrs Rose Lang for typing the text and coping with the continual additions necessary in this fast-moving field and Mrs Jane Templeman for her outstanding skill in preparing the illustrations.

David S. Latchman

Acknowledgements

I would like to thank all the colleagues, listed below, who have given permission for material from their papers to be reproduced in this book and have provided prints suitable for reproduction.

Figure 4.2, photograph kindly provided by Dr C. Wu from Zimarino and Wu, *Nature* **327**, 727 (1987), by permission of Macmillan Magazines Ltd. Figure 4.8, photograph kindly provided by Professor M. Beato from Willmann and Beato, *Nature* **324**, 688 (1986) by permission of Macmillan Magazines Ltd. Figures 5.9 and 5.12, photographs kindly provided by Dr R. L. Davis from Davis *et al.*, *Cell* **51**, 987 (1987) by permission of Cell Press. Figures 6.1 and 10.2, photographs kindly provided by Professor W.J. Gehring from Gehring, *Science* **236**, 1245 (1987) by permission of the American Association for the Advancement of Science. Figure 6.15, photograph kindly provided by Dr P. Holland from Holland and Hogan, *Nature* **321**, 251 (1986) by permission of Macmillan Magazines Ltd. Figures 6.17 and 6.18, photographs kindly provided by Dr R. Krumlauf from Graham *et al.*, *Cell* **57**, 367 (1989) by permission of Cell Press. Figure 8.8 redrawn from Redemann *et al.*, *Nature* **332**, 90 (1988) by kind permission of Dr H. Jackle and Macmillan Magazines Ltd. Figures 8.13 and 8.17, redrawn from Schwabe *et al.*, *Nature* **348**, 458 (1990) by kind permission of Dr D. Rhodes and Macmillan Magazines Ltd. Figure 8.20 , redrawn from Abel and Maniatis, *Nature* **341**, 24 (1989) by kind permission of Professor T. Maniatis and Macmillan Magazines Ltd. Figure 9.5, redrawn from Lin *et al.*, *Nature* **345**, 359 (1990) and Carey *et al.*, *Nature* **345**, 361 (1990) by kind permission of Professor M. Ptashne and Macmillan Magazines Ltd.

List of tables

CHAPTER ONE

DNA sequences and transcription factors

1.1 THE IMPORTANCE OF TRANSCRIPTION

The fundamental dogma of molecular biology is that DNA produces RNA which in turn produces protein. Hence if the genetic information that each individual inherits as DNA (the genotype) is to be converted into the proteins that produce the corresponding characteristics of the individual (the phenotype), it must first be converted into an RNA product. The process of transcription, whereby an RNA product is produced from the DNA, is therefore an essential element in gene expression. The failure of this process to occur will obviously render redundant all the other steps that follow the production of the initial RNA transcript in eukaryotes, such as RNA splicing, transport to the cytoplasm or translation into protein (reviews of these stages: see Nevins, 1983; Latchman, 1995).

The central role of transcription in the process of gene expression also renders it an attractive control point for regulating the expression of genes in particular cell types or in response to a particular signal. Indeed, it is now clear that in the vast majority of cases where a protein is produced only in a particular tissue or in response to a particular signal this is achieved by control processes that ensure its corresponding gene is transcribed only in that tissue or in response to such a signal (reviews: see Darnell, 1982; Latchman, 1995). For example, the genes encoding the immunoglobulin heavy and light chains of the antibody molecule are transcribed at high level only in the antibody-producing B cells (Gillis *et al.*, 1983), whilst the increase in somatostatin production in response to treatment of cells with cyclic AMP is mediated by increased transcription of the corresponding gene (Montminy *et al.*, 1986). Therefore, while post-transcriptional regulation affecting, for example, RNA splicing or stability plays some role in the regulation of gene expression (reviews: Peltz *et al.*, 1991; McKeown, 1992), the major control point lies at the level of transcription.

1.2 DNA SEQUENCE ELEMENTS

1.2.1 The gene promoter

The central role of transcription, both in the basic process of gene expression and its regulation in particular tissues, has led to considerable study of this process. Initially such studies focused on the nature of the DNA sequences within individual genes that were essential for either basal or regulated gene expression. In prokaryotes such sequences are found immediately upstream of the start site of transcription and form part of the promoter, directing expression of the genes. Sequences found at this position include elements found in all genes involved in the basic process of transcription itself and those found in a more limited number of genes which mediate their response to a particular signal (Miller and Reznikoff, 1980; Ptashne, 1986; Travers, 1993).

Early studies of cloned eukaryotic genes therefore concentrated on the region immediately upstream of the transcribed region where, by analogy, sequences involved in transcription and its regulation should be located. Putative regulatory sequences were identified by comparison between different genes, and the conclusions reached in this way were confirmed either by destroying these sequences by deletion or mutation or by transferring them to another gene in an attempt to alter its pattern of regulation.

This work, carried out on a number of different genes encoding specific proteins, identified many short sequence elements involved in transcriptional control (reviews: see Davidson *et al.*, 1983; Jones *et al.*, 1988). The elements of this type present in two typical examples; the human gene encoding the 70-kDa heat-inducible (heat-shock) protein (Williams *et al.*, 1989) and the human metallothionein IIA gene (Lee *et al.*, 1987a) are illustrated in Figure 1.1.

Comparisons of these and many other genes revealed that, as in bacteria, their upstream regions contain two types of elements: firstly, sequences found in very many genes exhibiting distinct patterns of regulation which are likely to be involved in the basic process of tran-

Figure 1.1 Transcriptional control elements upstream of the transcriptional start site in the human genes encoding *hsp70* (a) and methallothionein IIA (b). The TATA, Sp1 and CCAAT boxes bind factors that are involved in constitutive transcription (see also Chapter 3), whilst the glucocorticoid response element (GRE), metal response element (MRE), heat-shock element (HSE), and AP1 and AP2 sites bind factors involved in the induction of gene expression in response to specific stimuli (see also Chapter 4 and Section 7.2.1).

scription itself and, secondly, those found only in genes transcribed in a particular tissue or in response to a specific signal which are likely to produce this specific pattern of expression. These will be discussed in turn.

1.2.2 Sequences involved in the basic process of transcription

Although they are regulated very differently, the *hsp70* and metallothionein genes both contain a TATA box. This is an AT-rich sequence (consensus TATAA/TAA/T) which is found about 30 base pairs upstream of the transcriptional start site in many but not all genes. Mutagenesis or relocation of this sequence has shown that it plays an essential role in accurately positioning the start site of transcription (Breathnach and Chambon, 1981). The region of the gene bracketed by the TATA box and the site of transcriptional initiation (the Cap site) has been operationally defined as the gene promoter (Goodwin *et al.*, 1990). It is likely that this region binds several proteins essential for transcription, as well as RNA polymerase II itself, which is the enzyme responsible for transcribing protein-coding genes (Sentenac, 1985).

Although the TATA box is found in most eukaryotic genes, it is absent in some genes, notably housekeeping genes expressed in all tissues and in some tissue-specific genes (review: Weis and Reinberg, 1992). In these promoters, a sequence known as the initiator element which is located over the start site of transcription itself appears to play a critical role in determining the initiation point and acts as a minimal promoter capable of producing basal levels of transcription.

In promoters that contain a TATA box and in those that lack it, the low activity of the promoter itself is dramatically increased by other elements located upstream of the promoter. These elements are found in a wide variety of genes with different patterns of expression, indicating that they play a role in stimulating the constitutive activity of promoters. Thus inspection of the *hsp70* and metallothionein IIA genes reveals that both contain one or more copies of a GC-rich sequence known as the Sp1 box, which is found upstream of the promoter in many genes both with and without TATA boxes (review: Dynan and Tjian, 1985).

In addition the *hsp70* promoter but not the metallothionein promoter contains another sequence, the CCAAT box, which is also found in many genes with disparate patterns of regulation. Both the CCAAT box and the Sp1 box are typically found upstream of the TATA box as in the metallothionein and *hsp70* genes. Some genes, as in the case of *hsp70*, may have both of these elements, whereas others, such as the metal-lothionein gene, have single or multiple copies of one or the other (review: McKnight and Tjian, 1986). In every case, however, these elements are essential for transcription of the genes, and their elimination by deletion or mutation abolishes transcription (McKnight and Kingsbury, 1982; McKnight *et al.*, 1984). Hence these sequences play an essential role in efficient transcription of the gene and have been termed upstream promoter elements (UPE) (Goodwin *et al.*, 1990). The role of the promoter and UPE sequences, and of the protein factors that bind to them are discussed further in Chapter 3.

1.2.3 Sequences involved in regulated transcription

Inspection of the *hsp70* promoter (Figure 1.1) reveals several other sequence elements which are shared with a much more limited number of other genes and which are interdigitated with the upstream promoter elements discussed above. Indeed, one of these, which is located approximately 90 bases upstream of the transcriptional start site, is shared only with other heat-shock genes whose transcription is increased in response to raised temperature. This suggests that this heat-shock consensus element may be essential for the regulated transcription of the *hsp70* gene in response to heat.

To prove this directly, however, it is necessary to transfer this sequence to a non-heat-inducible gene and show that this transfer renders the recipient gene heat inducible. Pelham (1982) successfully achieved this by linking the heat-shock consensus element to the non-heat-inducible thymidine kinase gene of the eukaryotic virus herpes simplex. This hybrid gene could be activated following its introduction into mammalian cells by raising the temperature (Figure 1.2). Hence

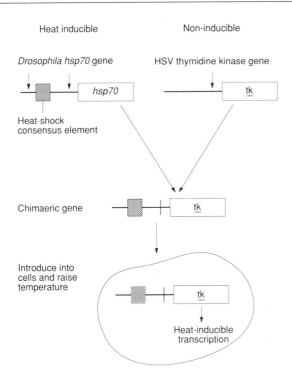

Figure 1.2 Demonstration that the heat-shock consensus element mediates heat inducibility. Transfer of this sequence to a gene (thymidine kinase) which is not normally inducible renders this gene heat inducible.

the heat-shock consensus element can confer heat inducibility on another gene, directly proving that its presence in the *hsp* gene promoters is responsible for their heat inducibility.

Moreover, although these experiments used a heat-shock consensus element taken from the *hsp70* gene of the fruit fly *Drosophila melanogaster*, the hybrid gene was introduced into mammalian cells. Not only does the successful functioning of the fly element in mammalian cells indicate that this process is evolutionarily conserved but it permits a further conclusion about the way in which the effect operates. Thus in the cold-blooded *Drosophila*, 37°C represents a thermally stressful temperature and the heat-shock response would normally be active at this temperature. The hybrid gene was inactive at 37°C in the mammalian cells, however, and was induced only at 42°C, the heat-shock temperature characteristic of the cell into which it was introduced. Hence this sequence does not act as a thermostat, set to go off at a particular temperature, since this would occur at the *Drosophila* heat-shock temperature (Figure 1.3a). Rather, this sequence must act by being recognized by a cellular protein which is activated only at an increased temperature characteristic of the mammalian cell heat-shock response (Figure 1.3b).

This experiment, therefore, not only directly proves the importance of the heat-shock response element in producing the heat inducibility of the *hsp70* gene but also shows that this sequence acts by binding a cellular protein that is activated in response to raised temperature. The binding of this transcription factor then activates transcription of the *hsp70* gene. The manner in which this factor activates transcription of the *hsp70* gene and the other heat-shock genes is discussed further in Section 4.2.

The presence of specific DNA sequences that can bind particular proteins will therefore confer on a specific gene the ability to respond to particular stimuli. Thus the lack of a heat-shock consensus element in the metallothionein IIA gene (Figure 1.1) means that this gene is not heat inducible. In contrast, however, this gene, unlike the *hsp70* gene, contains a glucocorticoid response element (GRE). Hence it can bind the complex of the glucocorticoid receptor and the hormone itself which forms following treatment of cells with glucocorticoid (review: Beato,

a) Thermostat DNA sequence

b) Thermostat protein

Figure 1.3 Predicted effects of placing the *Drosophila* heat-shock element (HSE) in a mammalian cell if the element acts as a thermostat detecting increased temperature directly (a) or if it acts by binding a protein that is activated by raised temperature (b). Note that only possibility (b) can account for the observation that the *Drosophila* HSE activates transcription in mammalian cells only at the mammalian heat-shock temperature of 42°C and not at the *Drosophila* heat-shock temperature of 37°C.

1989). Its transcription is therefore activated in response to glucocorticoid, whereas that of the *hsp70* gene is not (see Section 4.3). Similarly, only the metallothionein gene contains metal response elements (MRE), allowing it to be activated in response to treatment with heavy metals such as zinc and cadmium (Karin *et al.*, 1984). In contrast, both genes contain binding sites for the transcription factor AP2, which mediates gene activation in response to cyclic AMP and phorbol esters (Imagawa *et al.*, 1987). The manner in which the binding of specific transcription factors to different regulatory sequences modulates gene expression in response to specific inducing factors is discussed further in Chapter 4.

Similar DNA sequence elements in the promoters of tissue-specific genes play a critical role in producing their tissue-specific pattern of expression by binding transcription factors which are present in an active form only in a particular tissue where the gene will be activated. For example, the promoters of the immunoglobulin heavy- and light-chain genes contain a sequence known as the octamer motif (ATGCAAAT) (Parslow *et al.*, 1984; Mason *et al.*, 1985), which can confer B cell-specific expression on an unrelated promoter (Wirth *et al.*, 1987). This sequence acts by binding a transcription factor known as Oct-2 which is absent in most cell types but is expressed at high levels in immunoglobulin-producing B cells (Landolfi *et al.*, 1986; Staudt *et al.*, 1986) (see Section 5.2.2 for further details). Similarly, the related sequence ATGAATAA/T is found in two genes expressed specifically in the anterior pituitary gland, the prolactin gene and the growth hormone gene (Nelson *et al.*, 1988) and binds a transcription factor known as Pit-1, which is expressed only in the anterior pituitary (Ingraham *et al.*, 1988). If this short sequence is inserted upstream of a promoter, the gene is expressed only in pituitary cells. In contrast the octamer motif, which differs by only two bases, will direct expression only in B cells when inserted upstream of the same promoter (Elsholtz *et al.*, 1990; Figure 1.4). Hence small differences in control element sequences can produce radically different patterns of gene expression.

The role of transcription factors in producing tissue-specific gene expression is discussed further in Chapter 5.

1.2.4 Enhancers

One of the characteristic features of eukaryotic gene expression is the existence of sequence elements located at great distances from the start site of transcription which can influence the level of gene expression. These elements can be located upstream, downstream or within a transcription unit and function in either orientation relative to the

Figure 1.4 Linkage of the octamer-binding motif ATGCAAAT (1) and the related Pit-1-binding motif ATGAATAT (2) to the prolactin promoter and introduction into B cells and pituitary cells (a). In B cells, the octamer motif binds the B cell-specific octamer-binding protein Oct-2, whereas the Pit-1 motif fails to bind any protein. In contrast, in pituitary cells the Pit-1 motif binds the Pit-1 protein whereas the octamer motif fails to bind any protein (b). This results in only the octamer-containing construct 1 directing a high level of activity in B cells, whereas only construct 2 containing the Pit-1-binding site directs a high level of gene activity in pituitary cells (c). Data from Elsholtz *et al.* (1990).

start site of transcription (Figure 1.5). They act by increasing the activity of a promoter, although they lack promoter activity themselves and are hence referred to as enhancers (reviews: Hatzopoulos *et al.*, 1988; Muller *et al.*, 1988). Some enhancers are active in all tissues and increase the activity of a promoter in all cell types whilst others function as tissue-specific enhancers which activate a particular promoter only in a specific cell type. Thus the enhancer located in the intervening region of the immunoglobulin genes is active only in B cells (Gillis *et al.*, 1983) and the B cell-specific expression of the immunoglobulin genes is produced by the interaction of this enhancer and the immunoglobulin promoter which, as we have previously seen, is also B cell specific (Garcia *et al.*, 1986) (see Section 5.2 for further discussion).

As with promoter elements, enhancers contain multiple binding sites

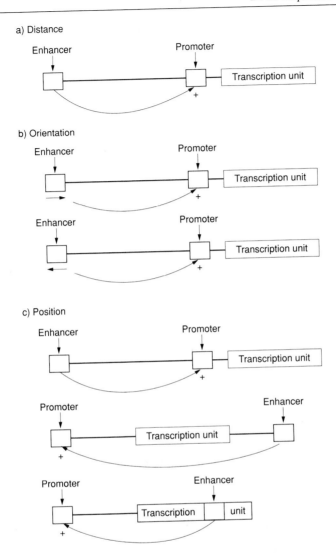

Figure 1.5 Characteristics of an enhancer element which can activate a promoter at a distance (a); in either orientation relative to the promoter (b); and when positioned upstream, downstream, or within a transcription unit (c).

for transcription factors which interact together (review: Dynan, 1989). In many cases these elements are identical to those contained immediately upstream of gene promoters. Thus the immunoglobulin heavy-chain enhancer contains a copy of the octamer sequence (Sen and Baltimore, 1986), which is also found in the immunoglobulin promoters (Section 1.2.3). Similarly multiple copies of the heat-shock consensus element are located far upstream of the start site in the *Xenopus hsp70* gene and function as a heat-inducible enhancer when transferred to another gene (Bienz and Pelham, 1986).

Enhancers therefore consist of sequence elements which are also present in similarly regulated promoters and may be found within the enhancer associated with other control elements or in multiple copies.

1.3 INTERACTION BETWEEN FACTORS BOUND AT VARIOUS SITES

The typical eukaryotic gene will therefore consist of up to four distinct transcriptional control elements (Figure 1.6). These are: firstly, the promoter itself; secondly, upstream promoter elements (UPE) located close to it which are required for efficient transcription in any cell type; thirdly, other elements adjacent to the promoter which are interdigitated with the UPEs and which activate the gene in particular tissues or in response to particular stimuli; and lastly, more distant enhancer elements which increase gene activity either in all tissues or in a regulated manner.

Such sequences act by binding positively acting transcription factors which then stimulate transcription (Figure 1.7a). Interestingly, however, although most sequences act in such a positive way, some sequences do appear to act in a negative manner to inhibit transcription. Such silencer elements have been defined in a number of genes including the cellular oncogene c-*myc* (Section 7.2.4) and those encoding

Figure 1.6 Structure of a typical gene with a TATA box-containing promoter, upstream promoter elements such as the CCAAT and Sp1 boxes, regulatory elements inducing expression in response to treatment with substances such as glucocorticoid (GRE) and cyclic AMP (CRE), and other elements within more distant enhancers. Note that, as discussed in the text and illustrated in Figure 1.1, the upstream promoter elements are often interdigitated with the regulatory elements, whilst the same regulatory elements can be found upstream of the promoter and in enhancers.

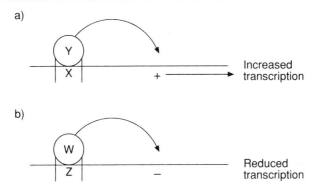

Figure 1.7 (a) A specific DNA sequence (X) can act to stimulate transcription by binding a positively acting factor (Y). (b) In contrast, binding of the negatively acting factor (W) to the DNA sequence (Z) inhibits transcription.

proteins such as growth hormone or collagen type II (review: Jackson, 1991). As with activating sequences, some silencer elements are constitutively active whilst others display cell type-specific activity. Thus, for example, the silencer in the gene encoding the T lymphocyte marker CD4 represses its expression in most T cells where CD4 is not expressed but is inactive in a subset of T cells, allowing these cells actively to express the CD4 protein (Sawada *et al.*, 1994). In many cases silencer elements have been shown to act by binding transcription factors which then act to reduce the rate of transcription (Figure 1.7b). The mechanisms by which this is achieved are discussed in Section 9.3.

Obviously the balance between positively and negatively acting transcription factors which bind to the regulatory regions of a particular gene will determine the rate of gene transcription in any particular situation. In some cases, binding of the RNA polymerase and associated factors to the promoter, and of other positive factors to the UPEs, will be sufficient for transcription to occur and the gene will be expressed constitutively. In other cases, however, such interactions will be insufficient and transcription of the gene will occur only in response to the binding, to another DNA sequence, of a factor that is activated in response to a particular stimulus or is present only in a particular tissue. These regulatory factors will then interact with the constitutive factors, allowing transcription to occur. Hence their binding will result in the observed tissue-specific or inducible pattern of gene expression.

Such interaction is well illustrated by the metallothionein IIA gene. As illustrated in Figure 1.1, this gene contains a binding site for the transcription factor AP1 which produces induction of gene expression in response to phorbol ester treatment (Angel *et al.*, 1987; Lee *et al.*, 1987b). The action of AP1 on the expression of the metallothionein

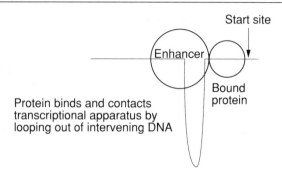

Figure 1.8 Contact between proteins bound at the promoter and those bound at a distant enhancer can be achieved by looping out of the intervening DNA.

gene is abolished, however, both by mutations in its binding site and by mutations in the adjacent Sp1 motif which prevent this motif binding its corresponding transcription factor Sp1 (Lee *et al.*, 1987a). Although these mutations in the Sp1 motif do not abolish AP1 binding, they do prevent its action, indicating that the inducible AP1 factor interacts with the constitutive Sp1 factor to activate transcription.

Clearly such interactions between bound transcription factors need not be confined to factors bound to regions adjacent to the promoter but can also involve the similar factors bound to more distant enhancers. It is likely that this is achieved by a looping out of the intervening DNA, allowing contact between factors bound at the promoter and those bound at the enhancer (Figure 1.8); for further discussion see Latchman (1995).

1.4 CONCLUSIONS

It is clear that both the process of transcription itself and its regulation in particular tissues or in response to particular signals are controlled by short DNA sequence elements located adjacent to the promoter or in enhancers. In turn, such sequences act by binding proteins which are either active constitutively or are present in an active form only in a specific tissue or following a specific inducing signal. Such DNA-bound transcription factors then interact with each other and the RNA polymerase itself to produce constitutive or regulated transcription. The nature of these factors, the manner in which they function and their role in different biological processes forms the subject of this book.

REFERENCES

Angel, P., Imagawa, M., Chiu, R., Stein, B., Imbra, R. J., Rahmsdorf, H. J., Jonat, C., Herrlich, P. and Karin, M. (1987). Phorbol ester-inducible genes contain a common *cis* element recognized by a TPA-modulated trans-acting factor. *Cell* **49**, 729–739.

Beato, M. (1989). Gene regulation by steroid hormones. *Cell* **56**, 335–344.

Bienz, M. and Pelham, H. R. B. (1986). Heat shock regulatory elements function as an inducible enhancer when linked to a heterologous promoter. *Cell* **45**, 753–760.

Breathnach, R. and Chambon, P. (1981). Organization and expression of eukaryotic split genes coding for proteins. *Annual Review of Biochemistry* **50**, 349–383.

Darnell, J. E. (1982). Variety in the level of gene control in eukaryotic cells. *Nature* **297**, 365–371.

Davidson, E. H., Jacobs, H. T. and Britten, R. J. (1983). Very short repeats and coordinate induction of genes. *Nature* **301**, 468–470.

Dynan, W. S. (1989). Modularity in promoters and enhancers. *Cell* **58**, 1–4.

Dynan, W. S. and Tjian, R. (1985). Control of eukaryotic messenger RNA synthesis by sequence-specific DNA-binding proteins. *Nature* **316**, 774–778.

Elshotz, H. P., Albert, V. R., Treacy, M. N. and Rosenfeld, M. G. (1990). A two-base change in a POU factor binding site switches pituitary-specific to lymphoid-specific gene expression. *Genes and Development* **4**, 43–51.

Garcia, J. V., Bich-Thuy, L., Stafford, J. and Queen, C. (1986). Synergism between immunoglobulin enhancers and promoters. *Nature* **322**, 383–385.

Gillis, S. D., Morrison, S. L., Oi, U. T. and Tonegawa, S. (1983). A tissue-specific transcription enhancer element is located in the major intron of a rearranged immunoglobulin heavy chain gene. *Cell* **33**, 717–728.

Goodwin, G. H., Partington, G. A. and Perkins, N. D. (1990). Sequence specific DNA binding proteins involved in gene transcription. In: *Chromosomes: Eukaryotic, Prokaryotic and Viral* (Adolph, K. W., ed.), vol. 1, pp. 31–85. Boca Raton, Florida: CRC Press.

Hatzopoulos, A. K., Schlokat, U. and Gruss, P. (1988). Enhancers and other *cis*-acting sequences. In: *Transcription and Splicing* (Hames, B. D. and Glover, D. M., eds), pp. 43–96. Oxford: IRL Press.

Imagawa, M., Chiu, R. and Karin, M. (1987). Transcription factor AP–2 mediates induction by two different signal-transduction pathways: protein kinase C and cAMP. *Cell* **51**, 251–260.

Ingraham, J. A., Chan, R., Mangalam, H. J., Elsholtz, H. P., Flynn, S. E., Linn, C. R., Simmons, D. M., Swanson, L. and Rosenfeld, M. G. (1988). A tissue specific factor containing a homeo domain specifies a pituitary phenotype. *Cell* **55**, 519–529.

Jackson, M. E. (1991). Negative regulation of eukaryotic transcription. *Journal of Cell Science* **100**, 1–7.

Jones, N. C., Rigby, P. W. J. and Ziff, E. B. (1988). Trans-acting protein factors and the regulation of eukaryotic transcription. *Genes and Development* **2**, 267–281.

Karin, M., Haslinger, A., Holtgreve, J., Richards, R. I., Krauter, P., Westphal, H. M. and Beato, M. (1984). Characterization of DNA sequences through which cadmium and glucocorticoid hormones induce metallothionein IIA gene. *Nature* **308**, 513–519.

Landolfi, N. F., Capra, D. J. and Tucker, P. W. (1986). Interaction of cell type-specific nuclear proteins with immunoglobulin VH promoter region sequences. *Nature* **323**, 548–551.

Latchman, D. S. (1995). Gene Regulation: A Eukaryotic Perspective, 2nd edn. London: Chapman and Hall.

Lee, W., Haslinger, A., Karin, M. and Tjian, R. (1987a). Activation of transcription by two factors that bind promoter and enhancer sequences of the human metallothionein gene and SV40. *Nature* **325**, 369–372.

Lee, W., Mitchell, P. and Tjian, R. (1987b). Purified transcription factor AP–1 interacts with TPA-inducible enhancer elements. *Cell* **49**, 741–752.

McKeown, M. (1992). Alternative mRNA splicing. *Annual Review of Cell Biology* **8**, 133–155.

McKnight, S. L. and Kingsbury, R. (1982). Transcriptional control signals of a eukaryotic protein coding gene. *Science* **217**, 316–324.

McKnight, S. and Tjian, R. (1986). Transcriptional selectivity of viral genes in mammalian cells. *Cell* **46**, 795–805.

McKnight, S. L., Kingsbury, R. C., Spence, A. and Smith, M. (1984). The distal transcription signals of the herpes virus tK gene share a common hexanucleotide control sequence. *Cell* **37**, 253–262.

Mason, J. O., Williams, G. T. and Neuberger, M. S. (1985). Transcription cell type specificity is conferred by an immunoglobulin VH gene promoter that includes a functional consensus sequence. *Cell* **41**, 479–487.

Miller, J. and Reznikoff, W.K. (eds) (1980). *The Operon*. New York: Cold Spring Harbor Laboratory.

Montminy, M. R., Sevarno, K. A., Wagner, J. A., Mondel, G. and Goodman, R. H. (1986). Identification of a cyclic AMP responsive element within the rat somatostatin gene. *Proceedings of the National Academy of Sciences USA* **86**, 6682–6686.

Muller, M. M., Gerster, T. and Schaffner, W. (1988) Enhancer sequences and the regulation of gene transcription. *European Journal of Biochemistry* **176**, 485–495.

Nelson, C. R., Albert, V. R., Elsholtz, H. P., Lu, L. E. W. and Rosenfeld, M. G. (1988). Activation of cell specific expression of rat growth hormone and prolactin genes by a common transcription factor. *Science* **239**, 1400–1505.

Nevins, J. R. (1983). The pathway of eukaryotic mRNA transcription. *Annual Review of Biochemistry* **52**, 441–446.

Parslow, T. G., Blair, D. L., Murphy, W. J. and Granner, D. K. (1984). Structure of the 5′ ends of immunoglobulin genes: a novel conserved sequence. *Proceedings of the National Academy of Sciences USA* **81**, 2650–2654.

Pelham, H. R. B. (1982). A regulatory upstream promoter element in the *Drosophila hsp70* heat-shock gene. *Cell* **30**, 517–528.

Peltz, S. W., Brewer, G., Bernstein, P. and Ross, J. (1991) Regulation of mRNA turnover in eukaryotic cells. *Critical Reviews in Eukaryotic Gene Expression* **1**, 99–126.

Ptashne, M. (1986). *A Genetic Switch*. Cambridge and Palo Alto, California: Cell Press and Blackwell Scientific Publications.

Sawada, S., Scarborough, J. D., Kileen, N. and Littman, D. R. (1994). A lineage-specific transcriptional silencer regulates CD4 gene expression during T lymphocyte development. *Cell* **77**, 917–929.

Sen, R. and Baltimore, D. (1986). Multiple nuclear factors interact with the immunoglobulin enhancer sequences. *Cell* **46**, 705–716.

Sentenac, A. (1985) Eukaryotic RNA polymerases. *CRC Critical Reviews in Biochemistry* **1**, 31–90.

Staudt, L. H., Singh, H., Sen, R., Wirth, T., Sharp, P. A. and Baltimore, D. (1986). A lymphoid-specific protein binding to the octamer motif of immunoglobulin genes. *Nature* **323**, 640–643.

Travers, A. (1993). *DNA – Protein Interactions*. London: Chapman and Hall.

Weis, L. and Reinberg, D. (1992). Transcription by RNA polymerase II initiator directed formation of transcription competent complexes. *FASEB Journal* **6**, 3300–3309.

Williams, G. T., McClanahan, T. K. and Morimoto, R. I. (1989). E1a transactivation of the human *hsp70* promoter is mediated through the basal transcriptional complex. *Molecular and Cellular Biology* **9**, 2574–2587.

Wirth, T., Staudt, L. and Baltimore, D. (1987). An octamer oligonucleotide upstream of a TATA motif is sufficient for lymphoid specific promoter activity. *Nature* **329**, 174–178.

CHAPTER TWO

Methods for studying transcription factors

2.1 INTRODUCTION

The explosion in the available information on transcription factors that has occurred in the past few years has arisen primarily because of the availability of new or improved methods for studying these factors. Initially such studies normally focus on identifying a factor that interacts with a particular DNA sequence and characterizing this interaction. Subsequently the protein identified in this way is further characterized and purified, and its corresponding gene isolated for further study. The methods involved in these two types of study will be considered in turn (for details of the methodologies involved see Latchman (1993)).

2.2 METHODS FOR STUDYING DNA–PROTEIN INTERACTIONS

2.2.1 DNA mobility shift assay

As discussed in Section 1.2, the initial stimulus to identify a transcription factor normally comes from the identification of a particular DNA sequence that confers a specific pattern of expression on a gene which carries it. The next step, therefore, after the identification of such a sequence, is to define the protein factors that bind to it. This can be

LIVERPOOL JOHN MOORES UNIVERSITY
LEARNING SERVICES

readily achieved by the DNA mobility shift or gel retardation assay (Fried and Crothers, 1981; Garner and Revzin, 1981).

This method relies on the obvious principle that a fragment of DNA to which a protein has bound will move more slowly in gel electrophoresis than the same DNA fragment without bound protein. The DNA mobility shift assay is carried out, therefore, by first radioactively labelling the specific DNA sequence whose protein-binding properties are being investigated. The labelled DNA is then incubated with a nuclear (Dignam *et al.*, 1983) or whole cell (Manley *et al.*, 1980) extract of cells prepared in such a way as to contain the DNA-binding proteins. In this way DNA–protein complexes are allowed to form. The complexes are then electrophoresed on a non-denaturing polyacrylamide gel and the position of the radioactive DNA visualized by autoradiography. If no protein has bound to the DNA, all the radioactive label will be at the bottom of the gel, whereas if a protein-DNA complex has formed radio-

Figure 2.1 DNA mobility shift or gel retardation assay. Binding of a cellular protein (B) to the radioactively labelled DNA causes it to move more slowly upon gel electrophoresis and hence results in the appearance of a retarded band upon autoradiography to detect the radioactive label.

active DNA to which the protein has bound will migrate more slowly and hence will be visualized near the top of the gel (Figure 2.1). (For methodological details see Dent and Latchman (1993)).

This technique can be used, therefore, to identify proteins that can bind to a particular DNA sequence in extracts prepared from specific cell types. Thus, for example, in the case of the octamer sequence discussed in Section 1.2.3, a single retarded band is detected when this sequence is mixed, for example with a fibroblast extract. In contrast, when an extract from immunoglobulin-producing B cells is used, two distinct retarded bands are seen (Figure 2.2). Since each band is produced by a distinct protein binding to the DNA, this indicates that, in addition to

Figure 2.2 DNA mobility shift assay using a radioactively labelled probe containing the binding site for octamer-binding proteins (ATGCAAAT) and extracts prepared from fibroblast cells (1) or B cells (2). Note that fibroblast cells contain only one protein, Oct-1 (01), capable of producing a retarded band, whereas B cells contain both Oct-1 and an additional tissue-specific protein, Oct-2 (02). The complexes formed by Oct-1 and Oct-2 on the labelled oligonucleotide in the absence of unlabelled oligonucleotide (track a) are readily removed by a 100-fold excess of unlabelled octamer oligonucleotide (track b). They are not removed, however, by a similar excess of a mutant octamer oligonucleotide (ATAATAAT) which is known not to bind octamer-binding proteins (track c) (Lenardo *et al.*, 1987) or of the binding site for the unrelated transcription factor Sp1 (track d) (Dynan and Tjian, 1983). This indicates that the retarded bands are produced by sequence-specific DNA-binding proteins which bind specifically to the octamer motif and not to mutant or unrelated motifs.

the ubiquitous octamer-binding protein Oct-1 which is present in most cell types, B cells also contains an additional octamer-binding protein, Oct-2, which is absent in many other cells (Landolfi *et al.*, 1986; Staudt *et al.*, 1986). The role of the Oct-2 protein in immunoglobulin gene expression is discussed further in Section 5.2.2.

As well as defining the proteins binding to a particular sequence, the DNA mobility shift assay can also be used to investigate the precise sequence specificity of this binding. This can be done by including in the binding reaction a large excess of a second DNA sequence that has not been labelled. If this DNA sequence can also bind the protein bound by the labelled DNA, it will do so. Moreover, binding to the unlabelled DNA will predominate since it is present in large excess. Hence the retarded band will not appear in the presence of the unlabelled competitor since only protein–DNA complexes containing labelled DNA are visualized on autoradiography (Figure 2.3b). In contrast, if the competitor cannot bind the same sequence as the labelled DNA, the complex with the labelled DNA will form and the labelled band will be visualized as before (Figure 2.3c).

Thus, by using competitor DNAs which contain the binding sites for previously described transcription factors, it can be established whether the protein detected in a particular mobility shift experiment is identical or related to any of these factors. Similarly, if competitor DNAs are used which differ in only one or a few bases from the original binding site, the effect of such base changes on the efficiency of the competitor DNA and hence on binding of the transcription factor can be assessed. Figure 2.2 illustrates an example of this type of competition approach, showing that the octamer-binding proteins Oct-1 and Oct-2 are efficiently competed away from the labelled octamer probe by an excess of identical unlabelled competitor but not by a competitor containing three base changes in this sequence which prevent binding (ATGCAAAT to ATAATAAT) (Lenardo *et al.*, 1987). Similarly no competition is observed, as expected, when the binding site of an unrelated transcription factor Sp1 (Dynan and Tjian, 1983) is used as the competitor DNA.

The DNA mobility shift assay therefore provides an excellent means of initially identifying a particular factor binding to a specific sequence and characterizing both its tissue distribution and sequence specificity.

2.2.2 DNAse I footprinting assay

Although the mobility shift assay provides a means of obtaining information on DNA–protein interaction, it cannot be used directly to localize the area of the contact between protein and DNA. For this purpose, the

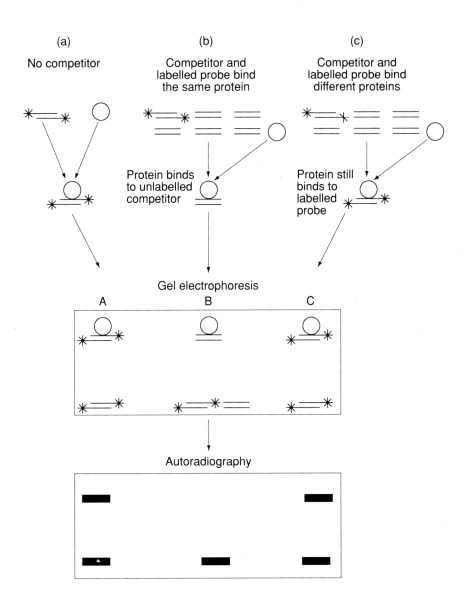

Figure 2.3 Use of unlabelled competitor DNAs in the DNA mobility shift assay. If an unlabelled DNA sequence is capable of binding the same protein as is bound by the labelled probe, it will do so (b) and the radioactive retarded band will not be observed, whereas if it cannot bind the same protein (c), the radioactive retarded band will form exactly as in the absence of competitor (a).

DNAse I footprint assay is used (Galas and Schmitz, 1978; Dynan and Tjian, 1983).

In this assay, DNA and protein are mixed as before, the DNA being labelled, however, only at the end of one strand of the double-stranded molecule. Following binding, the DNA is treated with a small amount of the enzyme deoxyribonuclease I (DNAse I), which will digest DNA. The digestion conditions are chosen, however, so that each molecule of DNA will be cut once or a very few times by the enzyme. Following digestion the bound protein is removed and the DNA fragments are separated by electrophoresis on a polyacrylamide gel capable of resolving DNA fragments differing in size by only one base. This produces a ladder of bands representing the products of DNAse I cutting either one or two or three

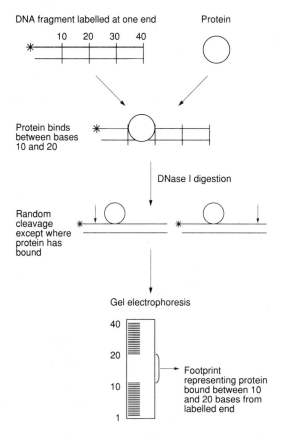

Figure 2.4 DNAse I footprinting assay. If a protein binds at a specific site within a DNA fragment labelled at one end, the region of DNA at which the protein binds will be protected from digestion with DNAse I. Hence this region will appear as a footprint in the ladder of bands produced by the DNA being cut at all other points by DNAse I.

or four etc. bases from the labelled end. Where a particular piece of the DNA has bound a protein, however, it will be protected from digestion and hence the bands corresponding to cleavage at these points will be absent. This will be visualized on electrophoresis as a blank area on the gel lacking labelled fragments and is referred to as the footprint of the protein (Figure 2.4). Similar labelling of the other strand of the DNA molecule will allow the interaction of the protein with the other strand of the DNA to be assessed. (For methodological details see Lakin, 1993.)

The footprinting technique therefore allows visualization of the inter-action of a particular factor with a specific piece of DNA. By using a sufficiently large piece of DNA, the binding of different proteins to different DNA sequences within the same fragment can be assessed. An analysis of this type is shown in Figure 2.5. This shows the footprints (A and B) produced by two cellular proteins binding to two distinct sequences within a region of the human immunodeficiency virus (HIV) control element which has an inhibiting effect on promoter activity (Orchard *et al.*, 1990). Interestingly, some insights into the topology of the DNA–protein interaction are also obtained in this experiment since bands adjacent to the protected region appear more intense in the presence of the protein. These regions of hypersensitivity to cutting are likely to represent a change in the structure of the DNA in this region when the protein has bound, rendering the DNA more susceptible to enzyme cleavage.

As with the mobility shift assay, unlabelled competitor sequences can be used to remove a particular footprint and determine its sequence specificity. In the HIV case illustrated in Figure 2.5, short DNA competi-tors containing the sequence of one or other of the footprinted areas were used specifically to remove each footprint without affecting the other, indicating that two distinct proteins produce the two footprints.

As well as footprinting using DNAse I, other footprinting techniques have been developed which rely on the protection of DNA which has bound protein from cleavage by other reagents that normally cleave the DNA. These include hydroxyl radical footprinting and phenanthroline–copper footprinting which, like DNAse I footprinting, rely on the ability of the reagents to cleave the DNA in a non-sequence-specific manner (for further details see Kreale, 1994; Papavassilou, 1995).

Of greater interest, however, is the technique of dimethyl sulphate (DMS) protection footprinting since it can provide information on the exact bases within the binding site that are contacted by the protein. Thus this method relies on the ability of DMS specifically to methylate guanine residues in the DNA. These methylated G residues can then be cleaved by exposure to piperidene, whereas no cleavage occurs at unmethylated G residues (Maxam and Gilbert, 1980). A protein bound

Figure 2.5 (a). DNAse I footprinting assay carried out on a region of the human immunodeficiency virus (HIV) control element. The two footprints (A and B) are not observed when no cell extract is added to the reaction (track A) but are seen when cellular extract is added in the absence of competitor (track B). Addition of unlabelled oligonucleotide competitor containing the DNA sequence of site A removes the site A footprint without affecting site B (track C), whilst an unlabelled oligonucleotide containing the site B DNA sequence has the opposite effect (track D). Both footprints are removed by a mixture of unlabelled site A and B oligonucleotides (track E). Arrows indicate the position of sites at which cleavage with DNAse I is enhanced in the presence of protein bound to an adjacent site, indicating the existence of conformational changes induced by protein binding. The track labelled G represents a marker track consisting of the same DNA fragment chemically cleaved at every guanine residue. (b). Position of sites A and B within the HIV control element. The arrow indicates the start site of transcription.

to the DNA will protect the guanine residues which it contacts from methylation and hence they will not be cleaved upon subsequent piperidene treatment. As in the other footprinting techniques, therefore, specific bands produced by such treatment of naked DNA are absent in the protein–DNA sample (see Lassar *et al.* (1989) for an example of this approach). Unlike the other methods, however, because cleavage occurs at specific guanine residues, this method identifies specific bases within the DNA which are contacted by the transcription factor protein (for methodological details see Lakin, 1993).

These footprinting techniques therefore offer an advance on the mobility shift assay, allowing a more precise visualization of the DNA–protein interaction.

2.2.3 Methylation interference assay

The pattern of DNA–protein interaction can also be studied in more detail with the methylation interference assay (Siebenlist and Gilbert, 1980). Like methylation protection, this method relies on the ability of DMS to methylate G residues, which can then be cleaved with piperidene. However, methylation interference is based on assessing whether the prior methylation of specific G residues in the target DNA affects subsequent protein binding.

Thus, the target DNA is first partially methylated with DMS so that, on average, only one G residue per DNA molecule is methylated (Maxam and Gilbert, 1980). Each individual DNA molecule will therefore contain some methylated G residues with the particular residues that are methylated being different in each molecule. These partially methylated DNAs are then used in a DNA mobility shift experiment with an appropriate cell extract containing the DNA-binding protein. After electrophoresis the band produced by the DNA which has bound protein and that produced by the DNA which has not are excised from the gel and treated with piperidine to cleave the DNA at the methylated G residues and not at unmethylated Gs. Clearly, if methylation of a particular G prevents protein binding then cleavage at this methylated G will be observed only in the DNA that failed to bind the protein. Conversely, if a particular G residue plays no role in binding, then cleavage at this G residue will be observed equally in both the DNA which bound the protein and that which failed to do so (Figure 2.6).

Figure 2.7 shows this type of analysis applied to the protein binding to site B within the negatively acting element in the HIV promoter (for the footprint produced by the binding of this protein see Figure 2.5). In this case the footprinted sequence was palindromic (Figure 2.7), suggesting that the DNA–protein interaction may involve similar binding to the

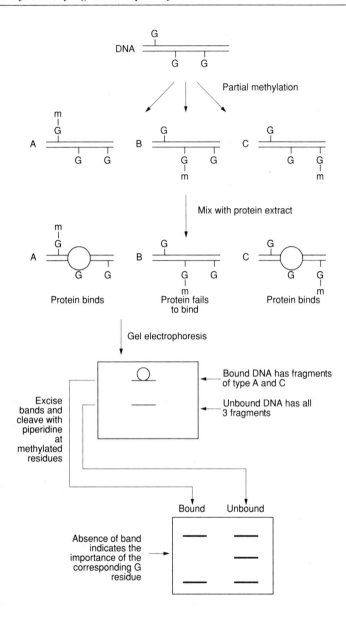

Figure 2.6 Methylation interference assay. Partially methylated DNA is used in a DNA mobility shift assay and both the DNA that has failed to bind protein and that which has bound protein and formed a retarded band are subsequently cleaved at methylated G residues with piperidine. If methylation at a specific G residue has no effect on protein binding (types A and C), the bound and unbound DNA will contain equal amounts of methylated G at this position. In contrast, if methylation at a particular G prevents binding of the protein (type B), only the unbound DNA will contain methylated G at this position.

a)

b)

-370   ** *   *   -320
TACACACCAGGGCCAGGGGGTCAGATATCCACTGACCTTTGGATGG

ATGTGTGGTCCCGGTCCCCAGTCTATAGGTGACTGGAAACCTACC
                                                        *   **

Figure 2.7 (a) Methylation interference assay applied to the DNA of site B in the HIV control element, as defined in the footprinting experiment shown in Figure 2.5. Both the upper (tracks A and B) and lower (tracks C and D) strands of the double-stranded DNA sequence were analysed. Tracks B and C show the methylation pattern of the unbound DNA that failed to bind protein, whereas tracks A and B show the methylation pattern of DNA that has bound protein. The arrows show G residues whose methylation is considerably lower in the bound compared with the unbound DNA and which are therefore critical for binding the specific cellular protein that interacts with this DNA sequence. (b) DNA sequence of site B. The extent of the footprint region is indicated by the square brackets and the critical G residues defined by the methylation interference assay in (a) are asterisked. Note the symmetrical pattern of critical G residues within the palindromic DNA sequence.

two halves of the palindrome. The methylation interference analysis of site B confirms this by showing that methylation of equivalent G residues in each half of the palindrome interferes with binding of the protein, indicating that these residues are critical for binding.

Although the DMS method only studies contacts of the protein with G residues, interference analysis can also be used to study the interaction of DNA-binding proteins with A residues in the binding site. This can be done either by methylating all purines to allow study of interference at A and G residues simultaneously (see, for example, Ares *et al.*, 1987) or by using diethylpyrocarbonate specifically to modify A residues (probably by carboxyethylation), rendering them susceptible to piperidine cleavage (see, for example, Sturm *et al.*, 1988). These techniques are of particular value when studying sequences such as the octamer motif in which there are relatively few G residues, hence limiting the information that can be obtained by studying interference at G residues alone (Sturm *et al.*, 1987; Baumruker *et al.*, 1988).

Chemical interference techniques can therefore be used to supplement footprinting methodologies and identify the precise DNA–protein interactions within the footprinted region.

2.2.4 *In vivo* footprinting assay

Although the methods described so far can provide considerable information about DNA–protein contacts, they all suffer from the deficiency that the DNA–protein interaction occurs *in vitro* when cell extract and DNA are mixed. Hence they indicate what factors can bind to the DNA rather then whether such factors actually do bind to DNA in the intact cell, where a particular factor may be sequestered in the cytoplasm or where its binding may be impeded by the association of DNA with other proteins such as histones.

These problems are overcome by the technique of *in vivo* footprinting, which is an extension of the *in vitro* DMS protection footprinting technique described in Section 2.2.2. Thus intact cells are freely permeable to DMS, which can therefore be used to methylate the DNA within its native chromatin structure in such cells. Exactly as in the *in vitro* technique, G residues to which a protein has bound will be protected from such methylation and will therefore not be cleaved when the DNA is subsequently isolated and treated with piperidene. Hence the bands produced by cleavage at these residues will be absent when the pattern produced by intact chromatin is compared with that produced by naked DNA (Figure 2.8).

Obviously the amounts of any specific DNA sequence obtained from total chromatin in this procedure are vanishingly small compared with when a cloned DNA fragment is used in the *in vitro* procedure. It is hence necessary to amplify the DNA of interest from within total chromatin by the polymerase chain reaction in order to obtain sufficient material for analysis by this method. When this is done, however, *in vivo* footprinting provides an excellent means for analysing DNA–protein contacts within intact cells *in vivo*, as well as determining the changes in such contacts that occur in response to specific treatments (for examples of this approach see Herrera *et al.*, 1989; Mueller and Wold, 1989).

Taken together, therefore, the three methods of DNA mobility shift, footprinting and methylation interference can provide considerable information on the nature of the interaction between a particular DNA sequence and a transcription factor. They serve as an essential prelude to a detailed study of the transcription factor itself.

Figure 2.8 In vivo footprinting using the methylation protection assay in which specific G residues are protected by bound protein (X) from methylation by dimethyl sulphate (DMS) treatment of intact cells. Hence, following DNA isolation, cleavage of methylated G residues with piperidine and subsequent amplification by the polymerase chain reaction (PCR), the band corresponding to cleavage at this protected residue will be absent. In contrast, cleavage at this position will be observed in naked DNA where no protein protects this residue from methylation.

2.3 METHODS FOR STUDYING THE TRANSCRIPTION FACTOR ITSELF

2.3.1 Protein purification

As discussed above, once a particular DNA sequence has been shown to be involved in transcriptional regulation, a number of techniques are available for characterizing the binding of transcription factors to this sequence. Although such studies can be carried out on crude cellular extracts containing the protein, ultimately they need to be supplemented by studies on the protein itself. This can be achieved by purifying the transcription factor from extracts of cells containing it. Unfortunately, however, conventional protein purification techniques such as conventional chromatography and high-pressure liquid chromatography (HPLC) result in the isolation of transcription factors at only 1–2% purity (Kadonaga and Tjian, 1986).

To overcome this problem and purify the transcription factor Sp1, Kadonaga and Tjian (1986) devised a method involving DNA affinity chromatography. In this method (Figure 2.9), a DNA sequence containing a high-affinity binding site for the transcription factor is synthesized and the individual molecules joined to form a multimeric molecule. This very high-affinity binding site is then coupled to an activated sepharose support on a column and total cellular protein passed down the column. The Sp1 protein binds specifically to its corresponding DNA sequence, whilst all other cellular proteins do not bind. The bound Sp1 can be eluted simply by raising the salt concentration. Two successive affinity chromatography steps of this type successfully resulted in the isolation of Sp1 at 90% purity, 30% of the Sp1 in the original extract being recovered, representing a 500–1000-fold purification (Kadonaga and Tjian, 1986).

Although this simple one-step method was successful in this case, it relies critically on the addition of exactly the right amount of non-specific DNA carrier to the cell extract. Thus, this added carrier acts to remove proteins that bind to DNA in a non-sequence-specific manner and which would hence bind non-specifically to the Sp1 affinity column and contaminate the resulting Sp1 preparation. This contamination will occur if too little carrier is added. If too much carrier is added, however, it will bind out the Sp1 since, like all sequence-specific proteins, Sp1 can bind with low affinity to any DNA sequence. Hence, in this case, no Sp1 will bind to the column itself (Figure 2.10).

To overcome this problem Rosenfeld and Kelley (1986) devised a method in which proteins capable of binding to DNA with high affinity in a non-sequence specific manner are removed before the extract is

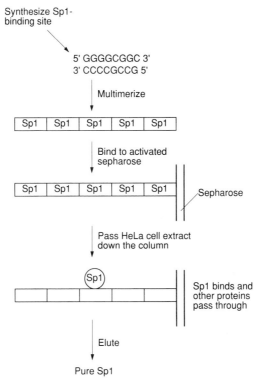

Figure 2.9 Purification of transcription factor Sp1 on an affinity column in which multiple copies of the DNA sequence binding Sp1 have been coupled to a sepharose support (Kadonaga and Tjian, 1986).

passed down the affinity column. To do this, the bulk of cellular protein was removed on a Biorex 70 high-capacity ion, exchange column and proteins that can bind to any DNA with high affinity were then removed on a cellulose column to which total bacterial DNA had been bound. Subsequently the remaining proteins, which had bound to non-sequence-specific DNA with only low affinity, were applied to a column containing a high-affinity binding site for transcription factor NF1 (Figure 2.11). NF1 bound to this site with high affinity and could be eluted in essentially pure form by raising the salt concentration (Table 2.1). It should be noted that in this and other purification procedures the fractions containing the transcription factor can readily be identified by carrying out a DNA mobility shift or footprinting assay with each fraction using the specific DNA binding site of the transcription factor.

The purified protein obtained in this way can obviously be used to characterize the protein, for example by determining its molecular weight or by raising an antibody to it to characterize its expression

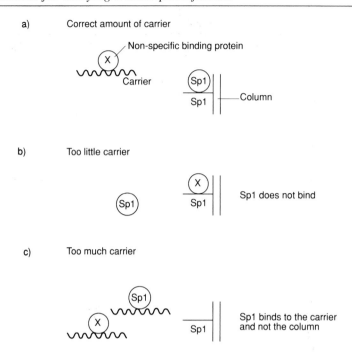

Figure 2.10 Consequences of adding different amounts of non-specific carrier DNA to the protein passing through the Sp1 affinity column. If the correct amount of non-specific carrier is added, it will bind proteins that interact with DNA in a non-sequence-specific manner, allowing Sp1 to bind to the column (a). However, addition of too little carrier will result in non-sequence-specific proteins binding to the column, thereby preventing the binding of Sp1 (b), whereas in the presence of too much carrier both the non-specific proteins and Sp1 will bind to the carrier (c).

pattern in different cell types. Similarly the activity of the protein can be assessed by adding it to cellular extracts and assessing its effect on their ability to transcribe exogenously added DNA in an *in vitro* transcription assay.

Unfortunately, however, because of the low abundance of transcription factors in the cell, these purification procedures yield very small amounts of protein. For example, Treisman (1987) succeeded in purifying only 1.6 µg of the serum response factor starting with 2×10^{10} cells or 40 g of cells. Such difficulties clearly limit the experiments that can be done with purified material. Indeed, the primary use of purified factor in most cases has simply been to provide material to isolate the gene encoding the protein. This gene can then be expressed either *in vitro* or in bacteria to provide a far more abundant source of the corresponding protein than could be obtained from cells that naturally express it.

Table 2.1 Purification of transcription factor NF1 from HeLa cells

	Total protein (mg)	Specific binding of ^{32}P DNA (fmol/mg protein) $\times 10^{-3}$	Purification (fold)	Yield (%)
HeLa cell extract*	4590	3.1	1.0	100
Biorex 70 column	550	27.1	8.7	104
E. coli DNA cellulose	65.2	181	58.4	83
NF1 affinity matrix				
1st passage	2.1	4510	1455	67
2nd passage	1.1	7517	2425	57

* Prepared from 6×10^{10} cells or 120 g cells.

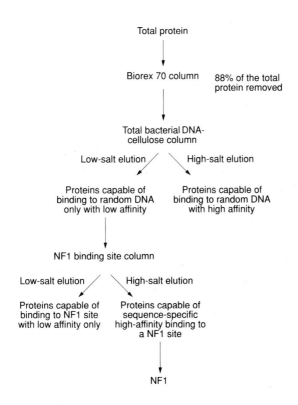

Figure 2.11 Purification of transcription factor NF-1 (Rosenfeld and Kelley, 1986). Following removal of most cellular proteins on a Biorex 70 ion-exchange column, proteins that bind to all DNA sequences with high affinity were removed on a bacterial DNA–cellulose column. Subsequent application of the remaining proteins to a column containing the NF1-binding site results in the purification of NF1, since it is the only protein that binds with low affinity to random DNA but with high affinity to an NF1 site.

2.3.2 Gene cloning

To isolate the gene encoding a particular transcription factor, a complementary DNA (cDNA) library is prepared from mRNA isolated from a cell type expressing the factor (for details of the methods used in preparing these libraries see Sambrook *et al.* (1989)). Some means is then required to identify a clone derived from the mRNA encoding the factor amongst all the other clones in the library. Two methods are normally used for this purpose.

(a) Use of oligonucleotide probes predicted from the protein sequence of the factor

If a particular transcription factor has been purified it is possible to obtain portions of its amino acid sequence. In turn, such sequences can be used to predict oligonucleotides containing DNA sequence, capable of encoding these protein fragments. Owing to the redundancy of the genetic code, whereby several different DNA codons can encode a particular amino acid, there will be multiple different oligonucleotides capable of encoding a particular amino acid sequence. All these possible oligonucleotides are synthesized chemically, made radioactive and used to screen the cDNA library. The oligonucleotide in the mixture which corresponds to the amino acid sequence of the transcription factor will hybridize to the corresponding sequence in a cDNA clone derived from mRNA encoding the factor. Hence such a clone can be readily identified in the cDNA library (Figure 2.12).

In cases where purified protein is available, as in those discussed in the previous section, this approach represents a relatively simple method for isolating cDNA clones. It has therefore been widely used to isolate cDNA clones corresponding to purified factors such as Sp1 (Kadonaga *et al.*, 1987; Figure 2.12), NF1 (Santoro *et al.*, 1988) and the serum response factor (Norman *et al.*, 1988) (for methodological details see Nicolas and Goodwin (1993)).

(b) Use of oligonucleotide probes derived from the DNA binding site of the factor

Although relatively simple, the use of oligonucleotides derived from protein sequence does require purified protein. As we have seen, purification of a transcription factor requires a vast quantity of cells and is technically difficult. Moreover, eventual determination of the partial

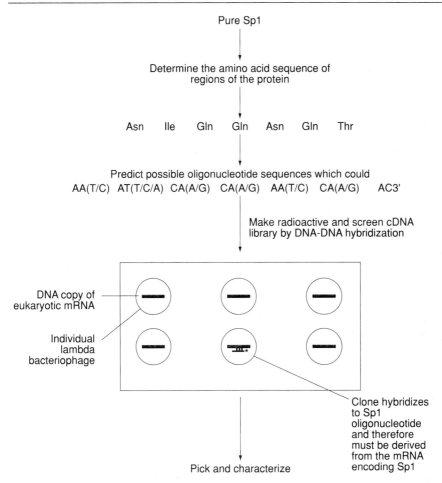

Pure Sp1

Determine the amino acid sequence of
regions of the protein

Asn   Ile   Gln   Gln   Asn   Gln   Thr

Predict possible oligonucleotide sequences which could
AA(T/C) AT(T/C/A) CA(A/G) CA(A/G) AA(T/C) CA(A/G)   AC3'

Make radioactive and screen cDNA
library by DNA-DNA hybridization

DNA copy of
eukaryotic mRNA

Individual
lambda
bacteriophage

Clone hybridizes
to Sp1
oligonucleotide
and therefore
must be derived
from the mRNA
encoding Sp1

Pick and characterize

Figure 2.12 Isolation of cDNA clones for the Sp1 transcription factor by screening with short oligonucleotides predicted from the protein sequence of Sp1. Because several different triplets of bases can code for any given amino acid, multiple oligonucleotides that contain every possible coding sequence are made. Positions at which these oligonucleotides differ from one another are indicated by the brackets containing more than one base.

amino acid sequence of the protein requires access to expensive protein-sequencing apparatus.

To bypass these problems, Singh *et al.* (1988) devised a procedure which is based on the fact that information is usually available about the specific DNA sequence to which a particular transcription factor binds. Hence a cDNA clone expressing the factor can be identified in a library by its ability to bind the appropriate DNA sequence. This method relies, therefore, on DNA–protein binding rather than DNA–DNA binding. Hence the library must be prepared in such a way that the cloned

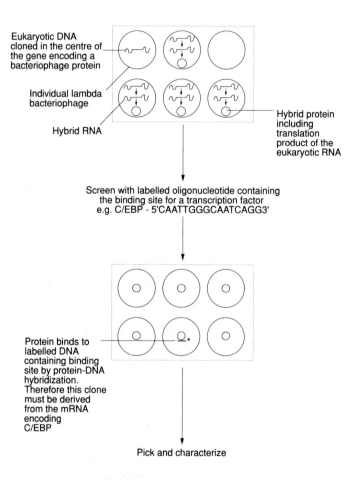

Figure 2.13 Isolation of cDNA clones for the C/EBP transcription factor by screening an expression library with a DNA probe containing the binding site for the factor.

cDNA inserts are translated by the bacteria into their corresponding proteins. This is normally achieved by inserting the cDNA into the coding region of the bacteriophage λ β-galactosidase gene, resulting in its translation as part of the bacteriophage protein (see Young and Davis (1983) for a further description of this bacteriophage vector and its other applications). The resulting fusion protein binds DNA with the same specificity as the original factor. Hence a cDNA clone encoding a particular factor can be identified in the library by screening with a radioactive oligonucleotide containing the binding site (Figure 2.13).

This technique has been used to isolate cDNA clones encoding several transcription factors such as the CCAAT box-binding factor C/EBP (Vinson *et al.*, 1988) and the octamer-binding proteins Oct-1 (Sturm *et al.*, 1988) and Oct-2 (Staudt *et al.*, 1988) (for methodological details see Cowell and Hurst (1993)).

The development of these two methods of screening with oligonucleotides derived from the protein sequence or from the binding site has therefore resulted in the isolation of cDNA clones corresponding to many transcription factors.

2.3.3 Use of cloned genes

This isolation of cDNA clones has in turn resulted in an explosion of information on these factors. Thus, once a clone has been isolated, its DNA sequence can be obtained, allowing prediction of the corresponding protein sequence and comparison with other factors. Similarly, the clone can be used to identify the mRNA encoding the protein and examine its expression in various tissues by Northern blotting, to study the structure of the gene itself within genomic DNA by Southern blotting, and as a probe to search for related genes expressed in other tissues or organisms.

Most importantly, however, the isolation of cDNA clones provides a means of obtaining large amounts of the corresponding protein for functional study. This can be achieved either by coupled *in vitro* transcription and translation (Figure 2.14a; Sturm *et al.*, 1988) or by expressing the gene in bacteria, either in the original expression vector used in the screening procedure (see Section 2.3.2b) or, more commonly, by subcloning the cDNA into a plasmid expression vector (Figure 2.14b Kadonaga *et al.*, 1987).

The protein produced in this way has similar activity to the natural protein, being capable of binding to DNA in footprinting or mobility shift assays (Kadonaga *et al.*, 1987) and of stimulating the transcription of appropriate DNAs containing its binding site when added to a cell-free transcription system (Mueller *et al.*, 1990).

Moreover, once a particular activity has been identified in a protein produced in this way, it is possible to analyse the features of the protein that produce this activity in a way that would not be possible using the factor purified from cells that normally express it. Thus, because the cDNA clone of the factor can readily be cut into fragments and each fragment expressed as a protein in isolation, particular features exhibited by the intact protein can easily be mapped to a particular region. Using the approach outlined in Figure 2.15, for example, it has proved possible to map the DNA-binding abilities of specific transcription

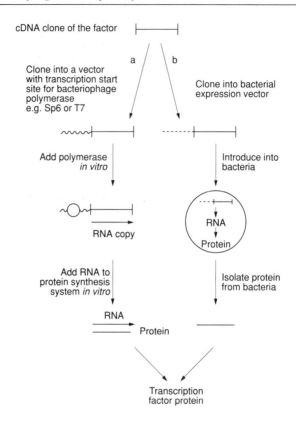

Figure 2.14 Methods of producing transcription factor protein from a cloned transcription factor cDNA. In the coupled *in vitro* transcription and translation method (a), the cDNA is cloned downstream of a promoter recognized by a bacteriophage polymerase and transcribed *in vitro* by addition of the appropriate polymerase. The resulting RNA is translated in an *in vitro* protein synthesis system to produce transcription factor protein. Alternatively (b) the cDNA can be cloned downstream of a prokaryotic promoter in a bacterial expression vector. Following introduction of this vector into bacteria, the bacteria will transcribe the cDNA into RNA and translate the RNA into protein, which can be isolated from the bacteria.

factors such as the octamer-binding proteins Oct-1 (Sturm *et al.*, 1988) and Oct-2 (Clerc *et al.*, 1988) to a specific short region of the protein. Once this has been done, particular bases in the DNA encoding the DNA-binding domain of the factor can then be mutated so as to alter its amino acid sequence, and the effect of these mutations on DNA binding can be assessed as before by expressing the mutant protein and measuring its ability to bind to DNA.

Approaches of this type have proved particularly valuable in defining DNA-binding motifs present in many factors and in analysing how

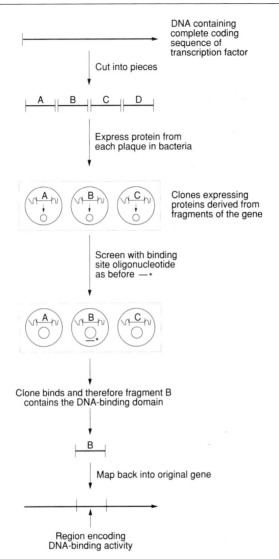

Figure 2.15 Mapping of the DNA-binding region of a transcription factor by testing the ability of different regions to bind to the appropriate DNA sequence when expressed in bacteria.

differences in the protein sequence of related factors define which DNA sequence they bind. This is discussed in Chapter 8.

One further piece of information to emerge from these studies is that the binding to DNA of a small fragment of the factor does not normally result in the activation of transcription. Thus a 60-amino-acid region of the yeast transcription factor GCN4 can bind to DNA in a sequence-specific manner but does not activate transcription of genes bearing its binding site (Hope and Struhl, 1986). Although DNA binding is necessary for transcription, therefore, it is not sufficient. This indicates that transcription factors have a modular structure in which the DNA-binding domain is distinct from another domain of the protein that mediates transcriptional activation.

The identification of the activation domain in a particular factor is complicated by the fact that DNA binding is necessary prior to activation. Hence the activation domain cannot be identified simply by expressing fragments of the protein and monitoring their activity. Rather, the various regions of the cDNA encoding the factor must each be linked to the region encoding the DNA-binding domain of another factor and the hybrid proteins produced. The ability of the hybrid factor to activate a target gene bearing the DNA-binding site of the factor supplying the

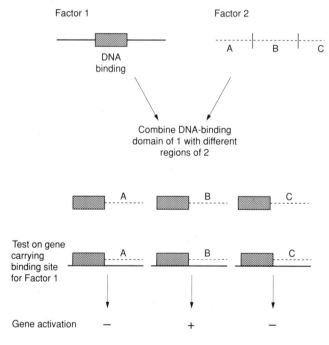

Figure 2.16 Domain-swapping experiment in which the activation domain of factor 2 is mapped by combining different regions of factor 2 with the DNA-binding domain of factor 1 and assaying the hybrid proteins for the ability to activate transcription of a gene containing the DNA-binding site of factor 1.

DNA-binding domain is then assessed (Figure 2.16). In these so-called 'domain-swap' experiments, binding of the factor to the appropriate DNA-binding site will be followed by gene activation only if the hybrid factor contains the region encoding the activation domain of the factor under test, allowing the activation domain to be identified.

Thus if another 60-amino-acid region of GCN4 distinct from the DNA-binding domain is linked to the DNA-binding domain of the bacterial Lex A protein, it can activate transcription in yeast from a gene containing a binding site for Lex A. This cannot be achieved by the Lex A DNA-binding domain or this region of GCN4 alone, indicating that this region of GCN4 contains the activation domain of the protein that can activate transcription following DNA binding and is distinct from the GCN4 protein DNA-binding domain (Hope and Struhl, 1986).

As with DNA-binding domains, the identification of activation domains and comparisons between the domains in different factors have provided considerable information on the nature of activation domains and the manner in which they function. This is discussed in Chapter 9.

2.4 CONCLUSIONS

This chapter has described a number of methods that allow investigation of the interaction of a transcription factor with DNA, its purification, gene cloning and dissection of its functional domains. The information obtained by the application of these procedures to particular factors is discussed in subsequent chapters.

REFERENCES

Ares, M. Jr, Chung, J.-S., Giglio, L. and Weiner, A. M. (1987). Distinct factors with Sp1 and NF-A specificities bind to adjacent functional elements of the human U2 snRNA gene enhancer. *Genes and Development* **1**, 808–817.

Baumruker, T., Sturm, R. and Herr, W. (1988). OBP 100 binds remarkably degenerate octamer motifs through specific interactions with flanking sequences. *Genes and Development* **2**, 1400–1413.

Clerc, R. G., Corcoran, L. M., LeBowitz, J. H., Baltimore, D. and Sharp, P. A. (1988). The B-cell specific Oct-2 protein contains POU box and homeo box type domains. *Genes and Development* **2**, 1570–1581.

Cowell, I .G. and Hurst, H. C. (1993). Cloning transcription factors from a cDNA expression library. In: *Transcription Factors: A Practical Approach* (Latchman, D. S., ed.) pp. 105–123. Oxford: IRL Press.

Dent, D. S. and Latchman, D. S. (1993). The DNA mobility shift assay. In: *Transcription Factors: A Practical Approach* (Latchman, D. S., ed.) pp. 1–26. Oxford: IRL Press.

Dignam, J. D., Lebovitz, R. M. and Roeder, R. G. (1983). Accurate transcription initiation by RNA polymerase II in a soluble extract from isolated mammalian nuclei. *Nucleic Acids Research* **11**, 1489–1575.

Dynan, W. S. and Tjian, R. (1983). The promoter specific transcription factor Sp1 binds to upstream sequences in the SV40 promoter. *Cell* **35**, 79–87.

Fried, M. and Crothers, D. M. (1981). Equilibria and kinetics of *lac* repressor–operator interactions by polyacrylamide gel electrophoresis. *Nucleic Acids Research* **9**, 6505–6525.

Galas, D. and Schmitz, A. (1978). DNAse footprinting: a simple method for the detection of protein–DNA binding specificity. *Nucleic Acids Research* **5**, 3157–3170.

Garner, M. M. and Revzin, A. (1981). A gel electrophoresis method for quantifying the binding of proteins to specific DNA regions: application to components of the *Escherichia coli* lactose operon regulatory system. *Nucleic Acids Research* **9**, 3047–3060.

Herrera, R. E., Shaw, P. E. and Nordheim, A. (1989). Occupation of the c-*fos* serum response element *in vivo* by a multi-protein complex is unaltered by growth factor induction. *Nature* **340**, 68–71.

Hope, I. A. and Struhl, K. (1986). Functional dissection of a eukaryotic transcriptional activator GCN4 of yeast. *Cell* **46**, 885–894.

Kadonaga, J. T. and Tjian, R. (1986). Affinity purification of sequence-specific DNA binding proteins. *Proceedings of the National Academy of Sciences USA* **83**, 5889–5893.

Kadonaga, J. T., Carner, K. R., Masiarz, F. R. and Tjian, R. (1987). Isolation of cDNA encoding the transcription factor Spl and functional analysis of the DNA binding domain. *Cell* **51**, 1079–1090.

Kreale, G. G. (ed.) (1994) *DNA–Protein Interactions*. New Jersey: Humana Press.

Lakin, N. D. (1993). Determination of DNA sequences that bind transcription factors by DNA footprinting. In: *Transcription Factors: A Practical Approach* (Latchman, D. S., ed.), pp 27–47. Oxford: IRL Press.

Landolfi, N. F., Capra, D. J. and Tucker, P. W. (1986). Interaction of cell type-specific nuclear proteins with immunoglobulin VH promoter region sequences. *Nature* **323**, 548–551.

Lassar, A. B., Buskin, J. N., Lockshun, D., Davis, R. L., Apone, S.,

Hanaschka, S. D. and Weintraub, H. (1989). Myo D is a sequence-specific DNA binding protein requiring a region of myc homology to bind to the muscle creatine kinase enhancer. *Cell* **58**, 823–831.

Latchman, D. S. (ed.) (1993). *Transcription Factors: A Practical Approach*. Oxford: IRL Press:

Lenardo, M., Pierce, J. W. and Baltimore, D. (1987). Protein binding sites in Ig gene enhancers determine transcriptional activity and inducibility. *Science* **236**, 1573–1577.

Manley, J. L., Fire, A., Cano, A., Sharp, P. A. and Gefter, M. L. (1980). DNA-dependent transcription of adenovirus genes in a soluble whole-cell extract. *Proceedings of the National Academy of Sciences USA* **77**, 3855–3859

Maxam, A. M. and Gilbert, W. (1980). Sequencing end labelled DNA with base-specific chemical cleavages. *Methods in Enzymology* **65**, 499–560.

Mueller, C. R., Macre, P. and Schibler, U. (1990). DBP a liver-enriched transcriptional activator is expressed late in ontogeny and its tissue specificity is determined post-transcriptionally. *Cell* **61**, 279–291.

Mueller, P. R. and Wold, B. (1989). *In vivo* footprinting of a muscle specific enhancer by ligation mediated PCR. *Science* **246**, 780–786.

Nicolas, R. H. and Goodwin G. H. (1993). Purification and cloning of transcription factors. In: *Transcription Factors: A Practical Approach* (Latchman, D. S., ed.), pp. 81–104. Oxford: IRL Press.

Norman, C., Runswick, M., Pollock, R. and Treisman, R. (1988). Isolation and properties of cDNA clones encoding SRF, a transcription factor that binds to the c-*fos* serum response element. *Cell* **55**, 989–1003.

Orchard, K., Perkins, N. D., Chapman, C., Harris, J., Emery, V., Goodwin, G., Latchman, D. S. and Collins, M. K. L. (1990). A novel T cell protein recognizes a palindromic element in the negative regulatory element of the HIV–1 LTR. *Journal of Virology* **64**, 3234–3239.

Papavassilou, A. G. (1995). Chemical nucleases as probes for studying DNA–protein interactions. *Biochemical Journal* **305**, 345–357.

Rosenfeld, P. J. and Kelley, T. J. (1986). Purification of nuclear factor 1 by DNA recognition site affinity chromatography. *Journal of Biological Chemistry* **261**, 1398–1408.

Sambrook, J., Fritsch, E. F. and Maniatis, T. (1989). *Molecular Cloning: A Laboratory Manual*. New York: Cold Spring Harbor Laboratory Press.

Santoro, C., Mermod, N., Andrews, P. C. and Tjian, R. (1988). A family of human CCAAT box binding proteins active in transcription and DNA replication: cloning and expression of multiple cDNAs. *Nature* **334**, 218–224.

Siebenlist, U. and Gilbert, W. (1980). Contacts between the RNA polymerase and an early promoter of phage T7. *Proceedings of the National Academy of Sciences USA* **77**, 122–126.

Singh, H., Le Bowitz, J. H., Baldwin, A. S. and Sharp, P. A. (1988). Molecular cloning of an enhancer binding protein: isolation by screening of an expression library with a recognition site DNA. *Cell* **52**, 415–429

Staudt, L. H., Singh, H., Sen, R., Wirth, T., Sharp, P. A. and Baltimore, D. (1986). A lymphoid-specific protein binding to the octamer motif of immunoglobulin genes. *Nature* **323**, 640–643.

Staudt, L. M., Clerc, R. G., Singh, H., Le Bowitz, J. H., Sharp, P. A. and Baltimore, D. (1988). Cloning of a lymphoid-specific cDNA encoding a protein binding the regulatory octamer DNA motif. *Science* **241**, 577–580.

Sturm, R., Baumruker, T., Franza, R. Jr and Herr, W. (1987). A 100-kD HeLa cell octamer binding protein (OBP 100) interacts differently with two separate octamer-related sequences within the SV40 enhancer. *Genes and Development* **1**, 1147–1160.

Sturm, R. A., Das, G. and Herr, W. (1988). The ubiquitous octamer-binding protein Oct-1 contains a POU domain with a homeobox subdomain. *Genes and Development* **2**, 1582–1599.

Treisman, R. (1987). Identification and purification of a polypeptide that binds to the c-*fos* serum response element. *EMBO Journal* **6**, 2711–2717.

Vinson, C. R., La Marco, K. L., Johnson, P. F., Landschulz, W. H. and McKnight, S. Z. (1988). *In situ* detection of sequence-specific DNA binding activity specified by a recombinant bacteriophage. *Genes and Development* **2**, 801–806.

Young, R. A. and Davis, R. W. (1983). Yeast polymerase II genes: isolation with antibody probes. *Science* **222**, 778–782.

CHAPTER THREE

Transcription factors and constitutive transcription

3.1 RNA POLYMERASES

Transcription involves the polymerization of ribonucleotide precursors into an RNA molecule using a DNA template. The enzymes that carry out this reaction are known as RNA polymerases. In eukaryotes three different enzymes of this type exist which are active on different sets of genes and can be distinguished on the basis of their different sensitivities to the fungal toxin α-amanitin (Table 3.1) (review: Sentenac, 1985). All the genes that code for proteins as well as those encoding some of the small nuclear RNAs involved in splicing are transcribed by RNA polymerase II. Because of the very wide variety of regulatory processes that these genes exhibit, much of this book is concerned with the interaction of different transcription factors with RNA polymerase II. Information is also available, however, on the interaction of such factors with RNA polymerase I, which transcribes the genes encoding the 28S, 18S and 5.8S ribosomal RNAs (Sommerville, 1984), and with RNA polymerase

Table 3.1 Eukaryotic RNA polymerases

	Genes transcribed	Sensitivity to α-amanitin
I	Ribosomal RNA (45S precursor of 28S, 18S and 5.8S rRNA)	Insensitive
II	All protein-coding genes, small nuclear RNAs U1, U2, U3, etc.	Very sensitive (inhibited 1 µg/ml)
III	Transfer RNA, 5S rRNA, small nuclear RNA U6, repeated DNA sequences: Alu, B1, B2, etc., 7SK, 7SL RNA	Moderately sensitive (inhibited 10 µg/ml)

III, which transcribes the transfer RNA and 5*S* ribosomal RNA genes (Cilberto *et al.*, 1983). These interactions are therefore discussed where appropriate.

All three RNA polymerases are large multi-subunit enzymes, RNA polymerase II, for example, having 8-10 subunits with sizes ranging from 240 to 10 kDa in size (Sentenac, 1985; Saltzman and Weinmann, 1989). Interestingly, cloning of the genes encoding the largest subunits of each of the three polymerases has revealed that they show homology to one another (Memet *et al.*, 1988). Similarly, chemical labelling experiments have indicated that the second largest subunit of each polymerase contains the active site of the enzyme (Riva *et al.*, 1987), whilst at least three smaller, non-catalytic subunits are shared by the three yeast polymerases (Woychik *et al.*, 1990). Such relationships evidently indicate a basic functional similarity between the three eukaryotic RNA polymerases and may also be indicative of a common evolutionary origin.

In addition to the conservation of function between the three eukaryotic enzymes, each individual enzyme exhibits strong conservation between different organisms. Thus the largest subunit of the mammalian RNA polymerase II enzyme is 75% homologous with that of the fruit fly *Drosophila* (Saltzman and Weinmann, 1989) and also shows homology to the equivalent enzymes in yeast (Memet *et al.*, 1988) and even *E. coli* (Ahearn *et al.*, 1987). All the eukaryotic RNA polymerase II enzymes contain a repeated region at the carboxyl end of the largest subunit, which contains multiple copies of the sequence Tyr-Ser-Pro-Thr-Ser-Pro-Ser. This sequence is unique to the largest subunit of RNA polymerase II and is present in multiple copies, being repeated 52 times in the mouse protein and 26 times in the yeast protein. As expected from its evolutionary conservation, the repeated region is essential for proper functioning of the enzyme and hence for cell viability, although its size can be reduced to some extent without affecting the activity of the enzyme (review: Young, 1991).

Interestingly, this repeated region serves as a site for phosphorylation and it is likely that such phosphorylation is critical for functioning of the polymerase (review: Drapkin *et al.*, 1993). Thus it appears that the dephosphorylated form of RNA polymerase II is the form that enters the basal transcriptional complex (see Section 3.2.4), whilst its phosphorylation triggers the start of transcriptional elongation to produce the RNA product. Such phosphorylation appears to be a means of regulating the rate of transcription with specific stimuli such as growth factors resulting in enhanced phosphorylation of the polymerase (Dubois *et al.*, 1994). In addition, as will be discussed in Section 9.2.3, it has been suggested that the repeated region serves as a target for the activation domains of specific transcription factors, which interact with it in order to increase the activity of the polymerase (Sigler, 1988). Hence this region appears to represent a critical target for cellular regulatory processes.

Whether this is the case or not, it is clear that, whilst the RNA polymerases possess the enzymatic activity necessary for transcription, they cannot function independently. Rather, transcription involves numerous transcription factors which must interact with the polymerase and with each other if transcription is to occur. The role of these factors is to organize a stable transcriptional complex containing the RNA polymerase and which is capable of repeated rounds of transcription.

3.2 THE STABLE TRANSCRIPTIONAL COMPLEX

3.2.1 Characteristics of the stable transcriptional complex

For all three eukaryotic polymerases, the initiation of transcription requires a multi-component complex containing the RNA polymerase and transcription factors. This complex has several characteristics that have led to it being referred to as a stable transcriptional complex (Brown, 1984). These are as follows:

1. The assembled complex is stable to treatment with low concentrations of specific detergents or to the presence of a competing DNA template, both of which would prevent its assembly.
2. The complex contains factors that are necessary for its assembly but not for transcription itself. These factors can therefore be dissociated once the complex has formed, without affecting transcription.
3. The complex of RNA polymerase and other factors necessary for transcription is stable through many rounds of transcription, resulting in the production of many RNA copies from the gene.

These characteristics are illustrated in Figure 3.1.

Much of the information on these complexes has been obtained by studying the relatively simple systems of RNA polymerases I and III and applying the information obtained to the RNA polymerase II situation. The stable complex formed by each of these enzymes will therefore be discussed in turn.

3.2.2 RNA polymerase I

The simplest complex known is found for the transcription of the ribosomal RNA genes by RNA polymerase I in *Acanthamoeba* (review:

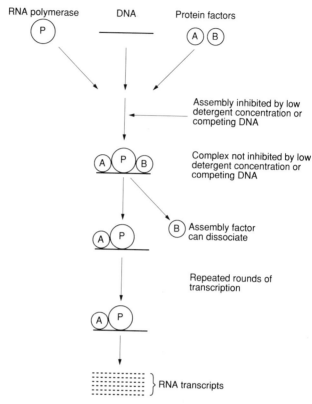

RNA polymerase   DNA   Protein factors

Figure 3.1 Stages in the formation of the stable transcriptional complex. The initial binding of the transcription factor (A) and the assembly factor (B) results in a meta-stable complex which can be dissociated by low levels of detergent or competing DNA. Following RNA polymerase binding, however, a stable complex is formed. This complex cannot be dissociated by low levels of detergent or competing DNA, is stable through multiple rounds of transcription and retains activity if the assembly factor (B) is removed.

Paule, 1990). In this organism only one transcription factor, known as TIF–1, is required for transcription by the polymerase. This factor binds to the ribosomal RNA promoter, protecting a region from 12 to 70 bases upstream of the transcriptional start site from DNAse I digestion. Subsequently the polymerase itself binds to the DNA just downstream of TIF–1, protecting a region between 18 and 52 bases upstream of the start site. Interestingly, binding of the polymerase is not dependent on the specific DNA sequence within this region since it can be replaced with a completely random sequence without affecting binding of the polymerase. Hence RNA polymerase is positioned on the promoter by protein–protein interaction with TIF–1, which has previously bound in a sequence-specific manner (Figure 3.2). When the RNA polymerase moves along the DNA transcribing the gene, TIF–1 remains bound at

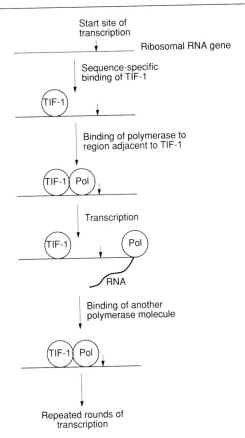

Figure 3.2 Transcription by *Acanthamoeba* RNA polymerase I involves the binding of transcription factor TIF-1 to a specific DNA sequence followed by binding of the polymerase in a non-sequence-specific manner to the DNA region adjacent to TIF-1. When the RNA polymerase moves away as it transcribes the gene, TIF-1 remains bound at the promoter, allowing another RNA polymerase molecule to bind and initiate a new round of transcription.

the promoter, allowing subsequent rounds of transcription to occur following binding of another polymerase molecule.

This system therefore represents a simple one in which a single factor is necessary for transcription and is active through multiple rounds of transcription. In vertebrate rRNA gene transcription, the situation is more complex, however, with an additional factor, UBF (upstream binding factor), also being involved (reviews: Reeder, 1992; Hernandez, 1993; Jacob, 1995). UBF binds specifically to the promoter and upstream elements of the rRNA genes and stimulates transcription (Jantzen *et al.*, 1992). This is achieved, however, by interaction with the vertebrate TIF–1 homologue, known as SL1. Thus, although a low basal rate of transcription is observed in the absence of UBF, no transcription is detectable unless SL1 is present. Unlike TIF–1, SL1 does not exhibit sequence-specific binding to the rRNA promoter. Hence UBF acts by binding to the DNA in a sequence-specific manner and facilitating the binding of SL1. Thus, whilst both SL1 and its homologue TIF–1 act as transcription factors necessary for polymerase I binding, UBF is an additional assembly factor required for binding of SL1 in vertebrates but not of TIF–1 in *Acanthamoeba*. This example therefore illustrates the distinction between factors required only for assembly of the complex or for binding of the polymerase and transcription itself (Figure 3.3).

3.2.3 RNA polymerase III

The different roles of transcription factors and assembly factors is also well illustrated by the RNA polymerase III system (for reviews see Cilberto *et al.*, 1983; Paule, 1990; Gabrielsen and Sentenac, 1991). Thus three different classes (I–III) of RNA polymerase III transcription unit exist, all of which require the essential factor TFIIIB for transcription (review: Hernandez, 1993). In the case of class I transcription units

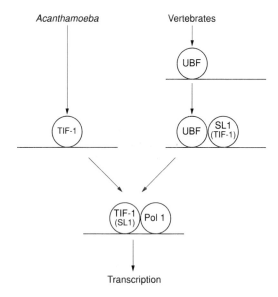

Figure 3.3 Comparison of ribosomal RNA gene transcription in *Acanthamoeba* and vertebrates; in vertebrates, transcription requires both the TIF-1 homologue SL1 and an additional assembly factor UBF whose prior binding is necessary for subsequent binding of SL1.

encoding the 5S rRNAs, transcription by RNA polymerase III requires the binding of three additional factors, TFIIIA, TFIIIB and TFIIIC. Although both TFIIIA and TFIIIC exhibit the ability to bind to 5S DNA in a sequence-specific manner, TFIIIB, like SL1, cannot do so unless TFIIIC has already bound. Once the complex of all these factors has formed and the RNA polymerase has bound, TFIIIA and TFIIIC can be removed and transcription continues with only TFIIIB and the polymerase bound to the DNA (Kassavetis *et al.*, 1990). Hence, like UBF, TFIIIA and TFIIIC are assembly factors required for the binding of the transcription factor TFIIIB. In turn, bound TFIIIB is recognized by the polymerase itself and transcription begins (Figure 3.4). As with RNA polymerase I, RNA polymerase III binds to the region of DNA adjacent to that which has bound the transcription factor, binding of the polymerase being independent of the DNA sequence in this region.

Although transcription of the class II RNA polymerase III transcription units, such as those encoding the tRNAs, is similar to that described for the 5S RNA genes, TFIIIA is not required. Rather, transcription is dependent only on TFIIIB and TFIIIC, with binding of TFIIIC being sufficient for subsequent binding of TFIIIB and the polymerase. Similarly, the class III RNA polymerase III transcription units, which have a

Figure 3.4 Binding of factors to the 5S RNA gene. Transcription requires the initial binding of the assembly factors TFIIIA and TFIIIC with subsequent binding of the transcription factor TFIIIB and of RNA polymerase III itself.

TATA box in the promoter (review: Sollner-Webb, 1988) which resembles that found in RNA polymerase II promoters (see Section 1.2.2), also require TFIIIB for transcription together with other accessory factors (for discussion see Hernandez, 1993).

The process of transcription by RNA polymerases I and III therefore involves the binding of a single transcription factor to the promoter, allowing subsequent binding of the RNA polymerase to an adjacent region of DNA. The transcription factor remains bound at the promoter as the polymerase moves down the DNA, allowing repeated binding of polymerase molecules and hence repeated rounds of transcription. Binding of the polymerase to the promoter requires prior binding of the transcription factor since the polymerase does not recognize a specific sequence in the promoter but rather makes protein–protein contact with the transcription factor and binds to the adjacent region of the DNA.

In different systems, however, different requirements exist for the binding of the transcription factor itself. Thus in the *Acanthamoeba* system, TIF–1 can bind to DNA in a sequence-specific manner and hence is the only factor required. In most other systems, this is not the case and the transcription factors do not bind to the DNA unless other assembly factors which exhibit sequence-specific DNA binding are present. Once the transcription factor has bound, these assembly factors can be removed, for example by detergent treatment, without affecting subsequent transcription. It is unclear, however, whether these factors do actually dissociate from the complex under normal conditions *in vivo* once the transcription factor has bound (for discussion see Paule, 1990). Whatever the case, the transcription factor itself remains bound at the promoter even after the polymerase has moved down the gene, allowing repeated binding of polymerase molecules and hence repeated rounds of transcription.

Although assembly factors play only an accessory role in transcription itself, they are essential if the complex is to assemble. Hence both assembly and transcription factors can be the target for processes that regulate the rate of transcription. Thus, whilst the high rate of polymerase III transcription in embryonal carcinoma cells is dependent on a high level of transcription factor TFIIIB (White *et al.*, 1989), the increase in transcription by this polymerase following adenovirus infection is due to an increase in the activity of assembly factor TFIIIC (Hoeffler *et al.*, 1988). Similarly, alterations in the level of TFIIIA during *Xenopus* development control the nature of the 5S rRNA genes which are transcribed at different developmental stages (Sakonju and Brown, 1982).

3.2.4 RNA polymerase II

As noted above, the very wide variety of regulatory events affecting the activity of genes transcribed by RNA polymerase II results in a bewildering array of transcription factors interacting with this enzyme and conferring particular patterns of regulation. These factors will be discussed in Chapters 4–7. Interestingly, however, even the basic transcriptional complex which is essential for any transcription by this enzyme contains far more components than is the case for the other RNA polymerases (reviews: Zawel and Reinberg, 1992; Conaway and Conaway, 1993; Buratowski, 1994).

One component of this complex, which has been intensively studied and which plays an essential role in RNA polymerase II-mediated transcription, is TFIID. In promoters containing a TATA box (see Section 1.2.2), TFIID binds to this element, protecting a region from 35 bases to 19 bases upstream of the start site of transcription in the human *hsp70* promoter, for example (review: Hernandez, 1993). The binding of TFIID to the TATA box or equivalent region is the earliest step in the formation of the stable transcriptional complex, such binding being facilitated by another factor, TFIIA (Figure 3.5a).

Interestingly, as TFIID is progressively purified, its requirement for TFIIA to aid its activity decreases. This is because, in less purified preparations and in the intact cell, TFIID is associated with a number of inhibitory factors such as DR1 and DR2 (review: Drapkin *et al.*, 1993) which act by preventing its binding to the DNA and/or its interaction with other components of the basal complex such as TFIIB (see below) (for further discussion of the role of Dr1, see Section 9.3.2). One role of TFIIA appears to be to bind to TFIID and overcome this inhibition, thereby stimulating the activity of TFIID. Hence, the need for TFIIA decreases as TFIID is purified away from these inhibitory factors, although it is likely to play a critical role in the intact cell. Indeed, the antagonism between positively and negatively acting factors in the assembly of the basal transcriptional complex may play a critical role in regulating the rate of transcription, representing a major target for activators of transcription (see Chapter 9, for further discussion of the mechanisms by which specific factors activate or inhibit transcription).

Once TFIID has bound to the DNA, another transcription factor TFIIB, joins the complex by binding to TFIID (Fgure 3.5b). This binding of TFIIB is an essential step in initiation complex formation since, as well as binding to TFIID, TFIIB can also bind to the RNA polymerase itself (review: Drapkin *et al.*, 1993). Hence it acts as a bridging factor, allowing the recruitment of RNA polymerase to the complex in association with another factor, TFIIF (Figure 3.5c). After polymerase binding, three other transcription factors, TFIIE, TFIIH and TFIIJ, rapidly

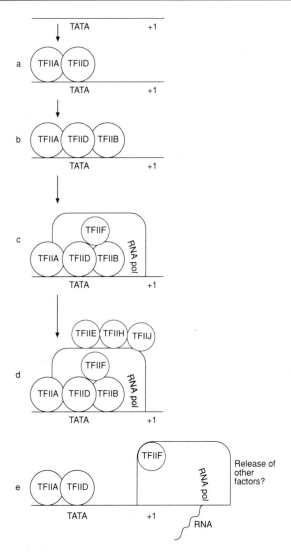

Figure 3.5 Stages in the assembly of the stable transcriptional complex for RNA polymerase II transcription. As the polymerase moves away from the promoter to transcribe the gene, TFIIF remains associated with it whilst TFIIA and TFIID remain bound at the TATA box, allowing the formation of a new stable complex and further rounds of transcription.

associate with the complex (Figure 3.5d). At this point, TFIIH, which has a kinase activity allowing it to phosphorylate other proteins, is believed to phosphorylate the RNA polymerase (Lu *et al.*, 1992). This converts it from the non-phosphorylated form, which joins the complex, to the phosphorylated form, which is capable of transcribing the DNA (see Section 3.1). Although this effect is likely to be of critical importance, it is not the only function of TFIIH. Thus, in addition, TFIIH also phosphorylates three other components of the complex TFIID, TFIIE and TFIIF (Ohkuma and Roeder, 1994) (for a review of TFIIH see Drapkin and Reinberg, 1994). Moreover, TFIIH also plays a critical role in the repair of damaged DNA, providing a possible link between the processes of DNA repair and transcription (review: Hanawalt, 1994), although different forms of TFIIH appear to be involved in the two processes (Suejstrup *et al.*, 1995).

The complex of the seven factors (TFIIA, B, D, E, F, H and J) and the polymerase is sufficient for transcription to occur. As the polymerase moves down the gene during this process, TFIIF remains associated with it whilst TFIIA and TFIID remain bound at the promoter and are capable of binding another molecule of polymerase, allowing repeated rounds of transcription as with the other polymerases (Figure 3.5e).

Although this step-by-step pathway of assembling the basal transcriptional complex is the generally accepted one, an alternative pathway has been proposed based on the finding that some RNA polymerase is found in solution already associated with TFIIB, TFIIF and TFIIH in the absence of DNA. It is possible, therefore, that, in some cases, following binding of TFIIA and TFIID to the promoter, this complex of RNA polymerase and associated factors may bind, resulting in a reduced number of steps being required for complex formation (for discussion see Carey, 1994; Koleske and Young, 1995).

Regardless of its method of assembly, however, it is clear that the basic stable transcriptional complex for RNA polymerase II requires a number of factors in addition to the polymerase itself and is therefore much more complex than that of RNA polymerase I or III.

3.2.5 TBP, the universal transcription factor?

Most of the transcription factors described in the previous sections were isolated by biochemical fractionation of cellular extracts and then shown to have a particular functional activity in modulating the rate of transcription when mixed with RNA polymerase and other subcellular fractions. When these factors were characterized in more detail by further fractionation and subsequent cloning, however, many of them were shown to consist of several different proteins which together are responsible for the properties ascribed to the original factor. Thus, although these factors have been dealt with for simplicity in the previous sections as single factors, they are in fact complexes of several different proteins; for example, TFIIE and TFIIF both contain two distinct proteins whilst TFIIH is a multi-protein complex with one of the component proteins having the kinase activity that results in phosphorylation of the RNA polymerase (see Section 3.2.4) (review: Buratowski; 1994).

This responsibility of one component of the complex for an activity formerly ascribed to the whole complex is seen most clearly in TFIID. Thus, TFIID is a multi-protein complex in which only one protein, known as TBP (TATA-binding protein), directs binding to the TATA box whilst the other components of the complex, known as TAFs (TBP-associated factors), do not bind directly to the TATA box and appear to allow TFIID to respond to stimulation by transcriptional activators (see Section 9.2.3).

Hence TBP plays a critical role in the transcription of TATA box-containing RNA polymerase II promoters by binding to the TATA box as the first step in assembly of the basal transcriptional complex. In view of this critical role, it is not surprising that TBP is one of the most highly conserved eukaryotic proteins (review: Hernandez, 1993). The structure of this protein has been defined by X-ray crystallography (Nikolov *et al.*, 1992) and shown to have a saddle structure in which the concave underside binds to DNA and the convex outer surface is accessible for interactions with other factors. Most interestingly, binding of TBP to the DNA deforms the DNA so that it follows the concave curve of the saddle (Figure 3.6). This change in DNA structure presumably facilitates transcription, either by unwinding the DNA or by promoting the binding of other components of the basal transcriptional complex

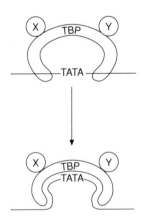

Figure 3.6 The saddle structure of TBP, as determined by X-ray crystallography, allows the concave surface to interact with the TATA box whilst the convex surface associates with other accessory transcription factors (X and Y). The initial binding induces bending of the DNA so that it follows the concave under-surface of the saddle.

(review: Klug, 1993). Similarly, TBP binds more efficiently to DNA which is already pre-bent (Parvin *et al.*, 1995), suggesting that factors which alter chromatin structure so as to produce such DNA bending may enhance TBP binding and thus assembly of the basal transcription complex.

Paradoxically, however, in view of its TATA box-binding ability, TBP also plays a critical role in the transcription of the subset of RNA polymerase II genes which do not contain a TATA box (see Section 1.2.2) (review: Weis and Reinberg, 1992). In this case, however, TBP does not bind to the DNA but is recruited to the promoter by another DNA-binding protein which binds to the initiator-element overlapping the transcriptional start site. TBP then binds to this initiator binding protein allowing the recruitment of TFIIB and the RNA polymerase itself as for promoters containing a TATA box. Hence TBP plays a

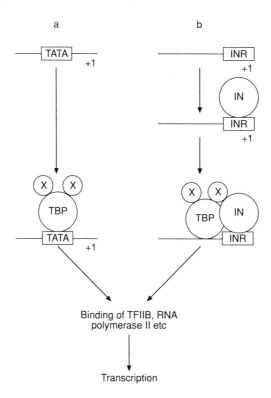

Figure 3.7 Transcription of promoters by RNA polymerase II involves the recruitment of TBP (and associated factors (X) forming the TFIID complex) to the promoter. This may be achieved by direct DNA binding to the TATA box where this is present (a) or by protein–protein interaction with a factor (IN) bound to the initiator element where the TATA box is absent (b).

critical role in the assembly of the transcription complex for RNA polymerase II, although it joins the complex by binding to DNA in the case of TATA box-containing promoters (Figure 3.7a) and is recruited by protein–protein interactions in the case of promoters that lack a TATA box (Figure 3.7b).

These findings have led to the suggestion that TBP represents the basic transcription factor for RNA polymerase II, paralleling the role of SL1 for RNA polymerase I and of TFIIIB for RNA polymerase III. This idea was supported by the amazing finding that TBP is actually also a component of both SL1 and TFIIIB (review: White and Jackson, 1992). Thus the SL1 factor is actually a complex of four factors, one of which is TBP (review: Sharp, 1992). Hence, when SL1 is recruited to the promoter by UBF (see Section 3.2.2), TBP is delivered to the DNA exactly as in the non TATA box-containing RNA polymerase II promoters where TBP is recruited by the prior binding of another protein to the initiator element.

Similarly, in the case of RNA polymerase III transcription where TBP is part of the multi-component TFIIIB complex (review: Rigby, 1993), TBP is delivered to class I polymerase III promoters by protein–protein interaction following the prior binding of TFIIIA and TFIIIC, and is delivered to class II polymerase III promoters by the prior binding of TFIIIC (Section 3.2.3) (Figure 3.8a). Interestingly, however, as noted in Section 3.2.3, the class III group of RNA polymerase III promoters contain a TATA box and hence in this case TBP can bind directly (Figure 3.8b). As in RNA polymerase II promoters, distinct mechanisms therefore ensure the recruitment of TBP to all RNA polymerase III promoters.

The similarities between the three RNA polymerases discussed in Section 3.1 are therefore paralleled by the involvement of a common factor, TBP, in transcription by all three RNA polymerases. In each case TBP forms a part of the multi-protein complexes that have been shown to be essential for transcription itself, binding via the TATA box or by protein–protein interactions with assembly factors (review: Struhl, 1994). In view of this, it has been suggested that TBP represents an evolutionarily ancient transcription factor preceding the division of the three RNA polymerases and having a universal and essential role in eukaryotic transcription (review: Hernandez, 1993). Indeed, recent evidence suggests that a TBP homologue is also found in the archaebacteria which constitute a separate kingdom distinct from eukaryotes and eubacteria (Rowlands *et al.*, 1994). Hence the existence of TBP appears to predate not only the divergence of the three RNA polymerases but also the divergence of the eukaryotic and archaebacterial kingdoms.

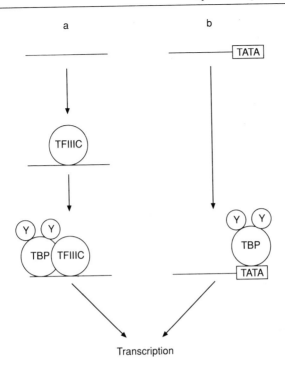

a   b

Transcription

Figure 3.8 Transcription of promoters by RNA polymerase III involves the recruitment of TBP (and associated factors (Y) forming the TFIIIB complex) to the promoter. This may be achieved by protein–protein interactions with TFIIIA and TFIIIC in the case of class I promoters, with TFIIIC alone in the case of class II promoters (a) or by direct DNA binding to the TATA box in class III promoters where the TATA box is present (b).

3.3 UPSTREAM PROMOTER ELEMENTS

3.3.1 Constitutive transcription factors

The complexity of the RNA polymerase II stable transcriptional complex is paralleled by the existence of numerous factors binding to elements upstream of the TATA box and increasing or decreasing the level of transcription (see Section 1.2). Many of these factors are active only in the presence of an inducing stimulus or in a specific tissue, thereby producing a specific pattern of inducible or tissue-specific gene expression. In addition, however, other factors which are constitutively active bind to specific upstream sequences. The binding of these factors and their interaction with the basal transcriptional complex results in increased levels of transcription in all tissues. This may be dependent on increased stability of the complex in the presence of these factors or to its increased activity. In the absence of such factors, therefore, the basal transcriptional complex can produce only a very low level of transcription and the binding of one or more of these factors is therefore necessary if significant levels of transcription are to occur. Two factors of this type will be discussed, namely those that bind to the Sp1 and CCAAT box elements found upstream of many promoters.

3.3.2 Sp1

The Sp1 box, which has the consensus sequence GGGCGG, was originally defined in the promoter of the eukaryotic virus SV40, which contains six copies of this motif, each of which binds the transcription factor Sp1 (McKnight and Tjian, 1986). Subsequently this sequence has been found in a variety of other promoters, such as the herpes simplex virus thymidine kinase promoter, the mouse dihydrofolate reductase promoter and the human metallothionein IIA promoter (reviews: Jones *et al.*, 1988; La Thangue and Rigby, 1988). As expected, the Sp1 protein is present in all cell types, two closely related Sp1 polypeptides of 105kDa and 95kDa having been purified, for example from HeLa cells. The addition of these proteins to an *in vitro* transcription reaction specifically stimulates the transcription of genes containing Sp1-binding sites, and the single gene encoding them has been cloned (Kadonaga *et al.*, 1987). Using the techniques described in Section 2.3.3, the DNA-binding activity of the protein encoded by the Sp1 gene has been shown to be dependent upon a region at the C-terminus which contains three so-called zinc finger motifs, also found in other transcription factors (review: Berg, 1992; also see Section 8.3.1 for further details of this DNA-binding motif). Similarly, the ability of the protein to activate transcription has been mapped to two glutamine-rich regions of the protein which act by interacting with TAFs that are associated with TBP within the TFIID complex (Courey and Tjian, 1988; Hoey *et al.*, 1993, see Sections 9.2.2 and 9.2.3 for further details of this activation motif and its mechanism of action). Hence the presence of the Sp1 protein in all cell types allows it to bind to its binding sites within specific gene promoters and results in their constitutive activation.

3.3.3 CCAAT box-binding proteins

Like the Sp1 motif, the CCAAT box has been found in a wide variety of promoters such as those of the thymidine kinase and *hsp70* genes and plays an essential role in their activity (reviews: McKnight and Tjian, 1986; Jones *et al.*, 1988). The CCAAT box binds a number of different proteins, some of which are expressed in all tissues whilst others are expressed in a tissue-specific manner (reviews: La Thangue and Rigby, 1988; Johnson and McKnight, 1989).

Initially, two distinct factors binding to this sequence were identified. One of these, CTF (CCAAT box transcription factor) consists of a family of polypeptides of 52-66 kDa in size (Rosenfeld and Kelley, 1986), which are encoded by multiple genes (Rupp *et al.*, 1990) with diversity being increased by alternative splicing of the primary transcripts (Santoro *et al.*, 1988). As well as their ability to stimulate transcription of genes containing a CCAAT box, these proteins can also stimulate DNA replication, CTF having been shown to be identical to nuclear factor 1, a protein that can stimulate the replication of adenovirus DNA *in vitro* (Jones *et al.*, 1987). Hence, like SV40 large T antigen and the octamer-binding protein Oct–1 (also referred to as nuclear factor III), this protein is capable of stimulating both transcription and DNA replication, and is hence referred to as CTF/NF1.

The second CCAAT box-binding protein to be defined, C/EBP, differs from CTF/NF1 in a number of properties such as its size (42 kDa), heat stability and the production of a different DNAse I footprint on the CCAAT box (Graves *et al.*, 1987; Johnson *et al.*, 1987). Most importantly, however, this protein differs from CTF/NF1 in its sequence specificity. Thus a C to G change in the first base of the CCAAT box impedes binding of CTF/NF1 but actually enhances the binding of C/EBP (Graves *et al.*, 1987). This enhanced binding of C/EBP reflects its ability to bind also to a core motif present in a wide variety of eukaryotic viral enhancers which has the consensus sequence TGTGGA/TA/TA/TG (Johnson *et al.*, 1987). The ability of C/EBP to bind with high affinity to these two distinct motifs led to its being termed CCAAT/enhancer binding protein (C/EBP) (for discussion see Johnson and McKnight (1989)).

The distinction between CTF/NF1 and C/EBP was confirmed when the genes encoding them were cloned and shown to be completely non-homologous with one another (Landschulz *et al.*, 1988; Santoro *et al.*, 1988). Indeed, although these two proteins can bind the identical DNA sequence in the CCAAT box, they do so using two completely different DNA binding motifs. Thus C/EBP contains the leucine zipper motif with adjacent basic DNA-binding domain common to a number of transcription factors, whilst CTF/NF1 contains a distinct DNA-binding motif first described in this factor (see Chapter 8 for discussion of these and other DNA-binding motifs). Hence sequence-specific binding to the same DNA sequence can be mediated by two distinct DNA-binding structures.

In addition to CTF/NF1 and C/EBP, numerous other CCAAT box-binding proteins have been reported in various tissues and cell types (e.g. Chodosh *et al.*, 1988a; Rajmondjean *et al.*, 1988). Of particular interest was the report by Chodosh *et al.* (1988a) of two multi-component CCAAT-binding activities. One of these, CP1, was formed by the combined activity of two factors, CP1A and CP1B, while the other, CP2, was formed by combining two other factors, CP2A and CP2B. Each of the four individual factors (CP1A, CP1B, CP2A and CP2B) or incorrect combinations (e.g. CP1A and CP2B) was unable to bind to the CCAAT box (Figure 3.9).

Most interestingly, the yeast catabolite derepression gene *cyc1* contains a sequence related to the CCAAT box (CCAAC) which is also bound only by a dimer of two proteins, HAP2 and HAP3, resulting in gene activation (Guarente *et al.*, 1984). Mixing of HeLa cell CP1A and yeast HAP2, or of CP1B and HAP3, regenerates a functional CCAAT box-binding activity (Chodosh *et al.* 1988b; Figure 3.10). Hence HAP2 and HAP3 are the yeast homologues of the human CP1B and CP1A proteins respectively. In agreement with this close relationship of the two pairs of proteins, mutation of the yeast *cyc1* gene sequence from CCAAC to CCAAT, creating a perfect CCAAT box, increases the binding affinity of the HAP2–HAP3 complex (Guarente *et al.*, 1984). This indicates, therefore, that a pair of proteins capable of interacting with one another and binding to the CCAAT box has been conserved from yeast to humans, such conservation having been sufficiently strong to allow correct interactions of yeast and human subunits.

The existence of multiple CCAAT box proteins clearly suggested that this sequence might play an important role in gene regulation apart from acting simply as an activator of constitutive gene regulation in all

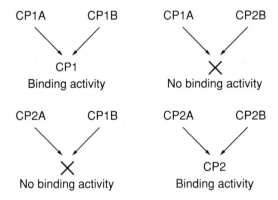

Figure 3.9 Generation of a functional CCAAT box-binding protein by mixing the appropriate subunits of CP1 or CP2 but not by mixing one subunit of CP1 with a subunit of CP2.

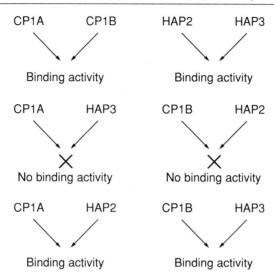

Figure 3.10 Generation of functional CCAAT box-binding protein by mixing subunits of CP1 with the yeast CCAAT box-binding subunits HAP2 and HAP3, indicating that HAP2 is the yeast homologue of CP1B and can therefore form a DNA-binding complex with CP1A, whilst HAP 3 is the yeast homologue of CP1A and can therefore form a DNA-binding complex with CP1B.

tissues, and evidence is available that this is the case. Thus, whilst the CTF/NF1 protein, like the Sp1 protein, is present in all tissues, C/EBP is expressed in the liver at much higher levels than in other tissues (Xanthopoulos *et al.*, 1989). Hence, while binding of CTF/NF1 to the CCAAT box is likely to play a role in constitutive expression of particular genes, C/EBP may activate the expression of genes expressed specifically in the liver. In agreement with this, the CCAAT box in several genes expressed specifically in the liver has been shown to play a role in their tissue-specific pattern of expression (e.g. Lichtsteiner *et al.*, 1987; Nerlov and Ziff, 1994). Interestingly, the disruption of C/EBP binding by a mutation in its binding site within one such liver-specific gene (that encoding factor IX) has recently been shown to be the cause of the human disease haemophilia B (Crossley and Brownlee, 1990). Similarly, C/EBP is specifically expressed during the terminal phase of adipocyte differentiation (review: Vasseur-Cognet and Lane, 1993) and when introduced into fibroblast cells can actually induce them to differentiate into adipocytes (Yeh *et al.*, 1995).

As well as functioning in a positive manner, the tissue-specific CCAAT box-binding factors can also be involved in tissue-specific gene regulation by acting negatively to prevent the binding of a constitutively expressed, positively acting factor. Thus, in the sea urchin, a factor known as the CCAAT displacement protein is present in embryonic

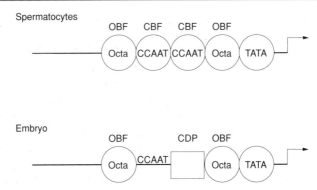

Figure 3.11 Binding of transcription factors to the sea urchin sperm histone H2B gene in spermatocytes where the gene is expressed and in embryos where it is not expressed. Note that the presence of a CCAAT displacement protein (CDP) in embryos prevents the binding of the positively acting CCAAT box-binding factor (CBF) to the two CCAAT boxes in the promoter. Expression of the H2B gene is obligately dependent on the binding of CBF and hence the gene is expressed only in sperm, where it is present, even though the otamer-binding factor (OBF) is bound to both copies of the octamer motif and TFIID is bound to the TATA box in both cell types.

tissues and binds to a sequence overlapping the CCAAT box in the sperm histone H2B gene. This binding prevents binding of the positively acting CCAAT box-binding protein and thereby prevents gene transcription (Figure 3.11). Hence the gene is expressed only in sperm when the displacement protein is absent and not in other tissues even though the positively acting CCAAT box-binding protein is present (Barberis *et al.*, 1987). A negative effect of C/EBP itself on the expression of human hepatitis B virus has also been reported, suggesting that the negative effect of CCAAT box-binding proteins may be widespread (Pei and Shih, 1990).

It is clear, therefore, that by binding a variety of constitutively expressed and tissue-specific proteins, the CCAAT box can contribute not only to constitutive expression of the genes which contain it but also to a variety of tissue-specific and developmental patterns of gene regulation.

3.4 CONCLUSIONS

The binding of each of the three eukaryotic RNA polymerases to appropriate gene promoters and subsequent transcription is dependent on the prior binding of a specific transcription factor to the promoter. Binding of the polymerase to the DNA adjacent to this factor occurs by recognition of the bound protein rather than by recognition of the specific DNA sequence in this region. In most cases, binding of the transcription factor itself requires the prior binding of other factors to the DNA. These assembly factors therefore play a critical role in the formation of the stable transcriptional complex but can be dissociated once the complex has formed without affecting its activity. In the case of RNA polymerase II transcription, either the stability of the complex or its activity is greatly affected by the binding of other proteins to sequences upstream of the promoter. Some of these proteins, such as Sp1 and CTF/NF1, which have been discussed in this chapter, are present in all tissues and therefore play a role in constitutive gene expression. Others are present or activated only in specific cell types or in response to a specific signal. The role of these proteins in the regulation of inducible or cell type-specific gene expression is discussed in Chapters 4 to 6.

REFERENCES

Ahearn, J. M. Jr, Bartolomei, M. S., West, M. L., Cisek, L. J. and Corden, J. L. (1987). Cloning and sequence analysis of the mouse genomic locus encoding the large subunit of RNA polymerase II. *Journal of Biological Chemistry* **262**, 10695–10705.

Barberis, A., Superti-Furga, G. and Busslinger, M. (1987). Mutually exclusive interaction of the CCAAT binding factor and of a displacement protein with overlapping sequences of a histone gene promoter. *Cell* **50**, 347–359.

Berg, J. M. (1992). Sp1 and the sub family of zinc finger proteins with guanine-rich binding sites. *Proceedings of the National Academy of Sciences USA* **89**, 11109–11110.

Brown, D. D. (1984). The role of stable complexes that repress and activate eukaryotic genes. *Cell* **37**, 359–365.

Buratowski, S. (1994). The basis of basal transcription by RNA polymerase II. *Cell* **77**, 1–3.

Carey, M. (1994). Transcription, simplifying the complex. *Nature* **368**, 402–403.

Chodosh, L. A., Baldwin, A. S., Carthew, R. W. and Sharp, P. A. (1988a). Human CCAAT-binding proteins have heterologous subunits. *Cell* **53**, 11–24.

Chodosh, L. A., Olesen, J., Hahn, S., Baldwin, A. S., Guarente, L. and Sharp, P. A. (1988b). A yeast and a human CCAAT-binding protein have heterologous subunits that are functionally interchangeable. *Cell* **53**, 25–35.

Cilberto, G., Castagnoli, L. and Cortese, R. (1983). Transcription by RNA polymerase III. *Current Topics in Developmental Biology* **18**, 59–88.

Conaway, R. C. and Conaway J. W. (1993). General initiation factors for RNA polymerase II. *Annual Review of Biochemistry* **62**, 161–190.

Courey, A. J. and Tjian, R. (1988). Analysis of Sp1 *in vivo* reveals multiple transcriptional domains including a novel glutamine-rich activation motif. *Cell* **55**, 887–898.

Crossley, M. and Brownlee, G. G. (1990). Disruption of a C/EBP binding site in the factor IX promoter is associated with haemophillia B. *Nature* **345**, 444–446.

Drapkin, R. and Reinberg, D. (1994). The multi-functional TFIIH complex and transcriptional control. *Trends in Biochemical Sciences* **19**, 504–508.

Drapkin, R., Merino, A. and Reinberg, D. (1993). Regulation of RNA polymerase II transcription. *Current opinion in Cell Biology* **5**, 469–476.

Dubois, M. F., Nguyer, V. T., Dahmus, M. E., Pages, G., Pouyssegur, J. and Bensaudo, O. (1994). Enhanced phosphorylation of the C-terminal domain of RNA polymerase II upon serum stimulation of quiescent cells: possible involvement of MAP kinases. *EMBO Journal* **13**, 4787–4797.

Gabrielsen, O. S. and Sentenac, A. (1991). RNA polymerase III (C) and its transcription factors. *Trends in Biochemical Sciences* **16**, 412–416.

Graves, B. J., Johnson, P. F. and McKnight, S. L. (1987). Homologous recognition of a promoter domain common to the MSV LTR and the HSV tk gene. *Cell* **44**, 565–576.

Guarente, L., Lalonde, B., Gifford, P. and Alani, E. (1984). Distinctly regulated tandem upstream activation sites mediate catabolite repression of the *cyc1* gene of *S. cerevisiae*. *Cell* **36**, 503–511.

Hanawalt, P. C. (1994). Transcription-coupled repair and human disease. *Science* **266**, 1957–1958.

Hernandez, N. (1993). TBP, a universal eukaryotic transcription factor. *Genes and Development* **7**, 1291–1308.

Hoeffler, W. K., Kovelman, R. and Roeder, R. G. (1988). Activation of transcription factor IIIC by the adenovirus E1A protein. *Cell* **53**, 907–920.

Hoey, T., Weinzierl, R. O. J., Gill, G., Chen, J. L., Dynlacht, B. D. and Tjian, R. (1993). Molecular cloning and functional analysis of *Drosophila* TAF110 reveals properties expected of coactivators. *Cell* **72**, 247–260.

Jacob, S. T. (1995). Regulation of ribosomal gene transcription. *Biochemical Journal* **306**, 617–626.

Jantzen, H-M., Ming Chow, A., King, D. S. and Tjian, R. (1992). Multiple domains of the RNA polymerase I activator hUBF interact with the TATA-binding protein hSL1 to mediate transcription. *Genes and Development* **6**, 1950–1963.

Johnson, P. F. and McKnight, S. L. (1989). Eukaryotic transcriptional regulatory proteins. *Annual Review of Biochemistry* **58**, 799–839.

Johnson, P. F., Landschulz, W. H., Graves, B. J. and McKnight, S. L. (1987). Identification of a rat liver nuclear protein that binds to the enhancer core element of three animal viruses. *Genes and Development* **1**, 133–146.

Jones, K. A., Kadonaga, J. T., Rosenfeld, P. J., Kelley, T. J. and Tjian, R. (1987). A cellular DNA binding protein that activates eukaryotic transcription and DNA replication. *Cell* **48**, 79–89.

Jones, N. C., Rigby, P. W. J. and Ziff, E. B. (1988). Trans-acting protein factors and the regulation of eukaryotic transcription. *Genes and Development* **2**, 267–281.

Kadonaga, J. T., Carner, K. R., Masiarz, F. R. and Tjian, R. (1987). Isolation of cDNA encoding the transcription factor Sp1 and functional analysis of the DNA binding domain. *Cell* **51**, 1079–1090.

Kassavetis, G. A., Braun, B. R., Nguyen, L. H. and Geiduschek, E. P. (1990). *S. cerevisiae* TFIIIB is the transcription initiation factor of RNA polymerase III while TFIIIA and TFIIC are assembly factors. *Cell* **60**, 235–245.

Klug, A. (1993). Transcription: opening the gateway. *Nature* **365**, 486–487.

Koleske, A. J. and Young, R. A. (1995). The RNA polymerase II holoenzyme and its implications for gene regulation. *Trends in Biochemical Sciences* **20**, 113–116.

Landshulz, W. H., Johnson, P. F., Adashi, E. Y., Graves, B. J. and McKnight, S. L. (1988). Isolation of a recombinant copy of the gene encoding C/EBP. *Genes and Development* **2**, 786–800.

La Thangue, N. B. and Rigby, P. W. J. (1988). Trans-acting protein factors and the regulation of eukaryotic transcription. In: *Transcription and Splicing* (Hames, B. D. and Glover, D., eds), pp. 3–42. Oxford: IRL Press.

Lichtsteiner, S., Wuari, J. and Schibler, U. (1987). The interplay of DNA-binding proteins on the promoter of the mouse albumin gene. *Cell* **51**, 963–973.

Lu, H., Zaiwel, L., Fisher, L., Egly, J. M. and Reinberg, D. (1992). Human general transcription factor IIH phosphorylates the C-terminal domain of RNA polymerase II. *Nature* **358**, 641–645.

McKnight, S. and Tjian, R. (1986). Transcriptional selectivity of viral genes in mammalian cells. *Cell* **46**, 795–805.

Memet, S., Saurn, W. and Sentenac, A. (1988). RNA polymerases B and C are more closely related to each other than to RNA polymerase A. *Journal of Biological Chemistry* **263**, 10048–10051.

Nerlov, C. and Ziff, E. B. (1994). Three levels of functional interaction determine the activity of the CCAAT enhancer binding protein-α on the serum albumin promoter. *Genes and Development* **8**, 350–362.

Nikolov, D. B., Hu, S. H., Lu, J., Gasch, A., Hoffmann, A., Horikoshi, M. Chu, N. H., Roeder, R. G. and Burley, S. K. (1992). Crystal structure of TFIID TATA-box binding protein. *Nature* **360**, 40–46.

Ohkuma, Y. and Roeder, R. G. (1994). Regulation of TFIIH ATPase and kinase activities by TFIIE during active initiation complex formation. *Nature* **368**, 160–163.

Parvin, J. D., McCormick, R. J., Sharp, P. A. and Fisher, D. E. (1995). Pre-bending of a promoter sequence enhances affinity for the TATA-binding factor. *Nature* **373**, 724–727.

Paule, M. R. (1990). In search of the single factor. *Nature* **344**, 819–820.

Pei, D. and Shih, C. (1990). Transcriptional activation and repression by cellular DNA binding protein C/EBP. *Journal of Virology* **64**, 1517–1522.

Rajmondjean, M., Cereghini, C. and Yaniv, M. (1988). Several distinct 'CCAAT' box binding proteins coexist in eukaryotic cells. *Proceedings of the National Academy of Sciences USA* **85**, 757–761.

Reeder, R. H. (1992). Regulation of transcription by RNA polymerase I. In: *Transcriptional Regulation* (McKnight, S. and Yamamoto, K.R., eds), pp. 315–348. Cold Spring Harbour, New York: Cold Spring Harbour Laboratory Press.

Rigby, P. W. J. (1993). Three in one and one in three: it all depends on TBP. *Cell* **72**, 7–10.

Riva, M., Schaffner, A. R., Sentenac, A., Hartmann, G. R., Mustner, A. A., Zaychikov, F. and Grachev, M. A. (1987) Active site labelling of the RNA polymerases A, B and C from yeast. *Journal of Biological Chemistry* **262**, 14377–14380.

Rosenfeld, P. J. and Kelley, T. J. (1986). Purification of nuclear factor 1 by DNA recognition site affinity chromatography. *Journal of Biological Chemistry* **261**, 1398–1408.

Rowlands, T., Baumann, P. and Jackson, S. P. (1994). The TATA-binding protein: a general transcription factor in eukaryotes and archaebacteria. *Science* **264**, 1326–1329.

Rupp, R. A. W., Kruse, U., Multhaup, G., Govel, U., Beyreuther, K. and Sippel, A. E. (1990). Chicken NF1/TGGCA proteins are encoded by at least three independent genes: NF1-A, NF1-B and NF1-C with homologues in mammalian genomes. *Nucleic Acids Research* **18**, 2607–2616.

Sakonju, S. and Brown, D. D. (1982). Contact points between a positive transcription factor and the *Xenopus* 5*S* RNA gene. *Cell* **31**, 395–405.

Saltzman, A. G. and Weinmann, R. (1989). Promoter specificity and modulation of RNA polymerase II transcription. *FASEB Journal* **3**, 1723–1733.

Santoro, C., Mermod, N., Andrews, P. C. and Tjian, R. (1988). A family of human CCAAT box binding proteins active in transcription and DNA replication: cloning and expression of multiple cDNAs. *Nature* **334**, 218–224.

Sentenac, A. (1985). Eukaryotic RNA polymerases. *CRC Critical Reviews in Biochemistry* **1**, 31–90.

Sharp, P. (1992). TATA-binding protein is a classless factor. *Cell* **68**, 819–821.

Sigler, P. B. (1988). Acid blobs and negative noodles. *Nature* **333**, 210–212.

Sollner-Webb, B. (1988) Surprises in RNA polymerase III transcription. *Cell* **52**, 153–154.

Sommerville, J. (1984). RNA polymerase I promoters and cellular transcription factors. *Nature* **310**, 189–190.

Struhl, K. (1994). Duality of TBP: the universal transcription factor. *Science* **263**, 1103–1104.

Suejstrup, J. Q., Wang, Z., Feaver, W. J., Wu, X., Bushnell, D. A., Donahue, T. F., Friedberg, E. C. and Kornberg, R. D. (1995). Different forms of TFIIH for transcription and DNA repair holo-TFIIH and a nucleosome excision repairosome. *Cell* **80**, 21–28.

Vasseur-Cognet, M. and Lane, M. B. (1993). Trans-acting factors involved in adipocyte differentiation. *Current Opinion in Genetics and Development* **3**, 238–245.

Weis, L. and Reinberg, D. (1992). Transcription by RNA polymerase II initiator directed formation of transcription-competent complexes. *FASEB Journal* **6**, 3300–3309.

White, R. J. and Jackson, S. P. (1992). The TATA binding protein: a central role in transcription by RNA polymerases I and III. *Trends in Genetics* **8**, 284–288.

White, R. J., Stott, P. and Rigby, P. W. J. (1989). Regulation of RNA polymerase III transcription in response to F9 embryonal carcinoma stem cell differentiation. *Cell* **59**, 1081–1092.

Woychik, N. A., Liao, S-M., Koldrieg, P. A. and Young, R. A. (1990). Subunits shared by eukaryotic RNA polymerases. *Genes and Development* **4**, 313–323.

Xanthopoulos, K. G., Mirkovitch, J., Decker, T., Kuo, C. F. and Darnell, J. E., Jr. (1989). Cell-specific transcriptional control of the mouse

DNA binding protein mC/EBP. *Proceedings of the National Academy of Sciences USA* **86**, 4117–4121.

Yeh, W-C., Cao, Z., Classon, M. and McKnight, S. L. (1995). Cascade regulation of terminal adipocyte differentiation by three members of the C/EBP family of leucine zipper proteins. *Genes and Development* **9**, 168–181.

Young, R. (1991). RNA polymerase II. *Annual Review of Biochemistry* **60**, 689–715.

Zawel, L and Reinberg, D. (1992). Advances in RNA polymerase II transcription. *Current Opinion in Cell Biology* **4**, 488–495.

CHAPTER FOUR

Transcription factors and inducible gene expression

4.1 INDUCIBLE GENE EXPRESSION

All cells, from bacteria to mammals, respond to various treatments by activating or repressing the expression of particular genes. As discussed in Section 1.2.3, genes that are activated in response to a specific treatment share a short DNA sequence in their promoters or enhancers whose transfer to another gene renders that gene inducible by the specific treatment. In turn, such sequences act by binding a specific transcription factor which becomes activated in response to the stimulus. Once activated, this factor interacts with the constitutive transcription factors discussed in Chapter 3, resulting in increased transcription of the gene.

A selection of DNA sequences that enable a gene to respond to a particular stimulus and the transcription factors which they bind is given in Table 4.1 (reviews: Maniatis *et al.*, 1987; Jones *et al.*, 1988). Rather than discuss each of these examples individually, we will focus on two cases of inducible gene expression. These two examples, the induction of gene activity by heat shock and by steroid hormones, have been chosen to illustrate the manner in which transcription factors can become active in response to a signal as well as how the active factor induces increased transcription of the target gene.

Table 4.1 Sequences that confer response to a particular stimulus

Consensus sequences	Response to	Protein factor	Gene containing sequences
CTNGAATNTT CTAGA	Heat	Heat-shock transcription factor	*hsp70, hsp83 hsp27*, etc.
T/G T/A CGTCA	Cyclic AMP	CREB/ATF	Somatostatin, fibronectin, α-gonadotrophin, c-*fos, hsp70*
TGAGTCAG	Phorbol esters	AP1	Metallothionein IIA, α_1-antitrypsin, collagenase
GATGTCCATATT AGGACATC	Growth factors in serum	Serum response factor	c-*fos, Xenopus* γ-actin
GGTACANNN TGTTCT	Glucocorticoid, progesterone	GR and PR receptors	Metallothionein IIA, tryptophan oxygenase, uteroglobin, lysozyme
AGGTCANNN TGACCT	Oestrogen	Oestrogen receptor	Ovalbumin, conalbumin, vitellogenin
TCAGGTCAT GACCTGA	Thyroid hormone, retinoic acid	TH and RA receptors	Growth hormone, myosin heavy chain
TGCGCCCGCC	Heavy metals	Mep-1	Metallothionein genes
AGTTTCNN TTTCNC/T	Interferon-α	Stat 1 Stat 2	Oligo A synthetase, guanylate-binding protein
TTNCNNNAA	Interferon-γ	Stat 1	Guanylate-binding protein, Fc γ receptor

4.2 HEAT-INDUCIBLE TRANSCRIPTION

4.2.1 The heat-shock factor

When cells from a variety of species, ranging from bacteria to mammals, are subjected to raised temperatures (heat shock) or other stresses such as anoxia, they respond by inducing increased transcription of a small number of genes known as the heat-shock genes (reviews: Morimoto, 1993; Parsell and Lindquist, 1993). As discussed in Section 1.2.3, these heat-inducible genes share a common DNA sequence which, when transferred to another gene, can render the second gene heat inducible. This sequence is known as the heatshock element (HSE). The manner in which a *Drosophila* HSE, when introduced into mammalian cells, functioned at the mammalian rather that the *Drosophila* heat-

Figure 4.1 Proteins binding to the promoter of the *hsp70* gene before (a) and after (b) heat shock.

shock temperature suggested that this sequence acted by binding a protein rather than by acting directly as a thermosensor (Figure 1.3).

Direct evidence that this was the case was provided by studying the proteins bound to the promoters of the *hsp* genes before and after heat shock. Thus, prior to heat shock, the TFIID complex (Sections 3.2.4 and 3.2.5) is bound to the TATA box and another transcription factor known as GAGA is bound upstream (Figure 4.1a). (Wu, 1985; Tsukiyama *et al.*, 1994).

Following heat-shock, however, an additional factor is observed which is bound to the HSE (Figure 4.1b). The amount of this factor bound to the HSE increased with the time of exposure to raised temperature and with the extent of temperature increase. Moreover, increased protein binding to the HSE was also observed following exposure to other agents that also induce the transcription of the heat-shock genes such as 2,4-dinitrophenol (Figure 4.2). Thus activation of the heat-shock genes, mediated by the HSE, is accompanied by the binding of a specific transcription factor to this DNA sequence. This factor, which has been variously referred to as the heat-shock transcription factor or heat-shock activator protein, is now generally known as the heat-shock factor (HSF).

Hence, prior to heat shock, the heat-shock genes are poised for transcription. Thus, whilst the bulk of cellular DNA is associated with histone proteins to form a tightly packed chromatin structure, the binding of the GAGA factor to the heat-shock gene promoters results in displacement of the histone-containing nucleosomes from the promoter region (Tsukiyama *et al.*, 1994; Van Holde, 1994). This opens up the chromatin and renders the promoter region exquisitely sensitive to digestion with the enzyme DNAse I (Figure 4.3). Although such a

HEAT SHOCK

Figure 4.2 Detection of HSF binding to the HSE 91 bases upstream (−91) of the start site for transcription in the *Drosophila hsp82* gene and protecting this region from digestion with exonuclease III. Note the increased binding of HSF with increasing time of exposure to heat shock or increased severity of heat shock. HSF binding is also induced by exposure to 2,4-dinitrophenol (DNP), which is known to induce transcription of the heat-shock genes.

DNAse I-hypersensitive site marks a gene as poised for transcription (review: Latchman, 1995), it is not in itself sufficient for transcription. This is achieved following heat shock by the binding of the HSF to the HSE. This factor then interacts with TFIID and other components of the basal transcription complex, resulting in the activation of transcription (for further discussion see Sections 9.2.3 and 9.2.4).

The critical role of the HSF in this process obviously begs the question of how this factor is activated in response to heat.

4.2.2 Activation of HSF by heat

If cells are heat-treated in the presence of cycloheximide, which is an inhibitor of protein synthesis, increased binding of HSF to the HSE is observed, exactly as in cells treated in the absence of the drug (Zimarino and Wu, 1987). This indicates that the observed binding of HSF following heat shock does not require *de novo* protein synthesis. Rather, this factor must pre-exist in non-heat-treated cells in an inactive form whose ability to bind to the HSE sequence in DNA is activated post-transla-

Heat-shock genes

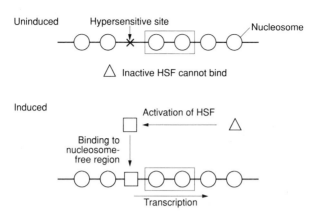

Figure 4.3 Activation of HSF by heat is followed by its binding to a pre-existing nucleosome-free region in the heat-shock gene promoters which is marked by a DNAse I-hypersensitive site. Binding of HSF then results in the activation of heat-shock gene transcription.

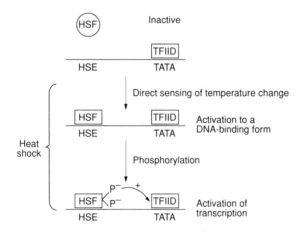

Figure 4.4 Stages in the activation of HSF in mammalian and *Drosophila* cells. Initial activation of HSF to a DNA-binding form following raised temperature is followed by its phosphorylation, which converts it to a form capable of activating transcription.

tionally by heat. In agreement with this, activation of HSF can also be observed following heat treatment of cell extracts *in vitro*, when new protein synthesis would not be possible (Larson *et al.*, 1988).

Analysis of the activation process using *in vitro* systems from human cells (Larson *et al.*, 1988) has indicated that it is a two-stage process (Figure 4.4). In the first stage, the HSF is activated to a form which can bind to DNA by an ATP-independent mechanism which is directly dependent on increased temperature. Subsequently this protein is further modified by phosphorylation, allowing it to activate transcription. Interestingly, the second of these two stages appears to be disrupted in murine erythroleukaemia (MEL) cells, in which heat shock results in increased binding of HSF to DNA but transcriptional activation of the heat-shock genes is not observed (Hensold *et al.*, 1990).

The activation of HSF into a form capable of binding DNA involves its conversion from a monomeric to a trimeric form which can bind to the HSE (review: Morimoto, 1993). The maintenance of the monomeric form of HSF prior to heat shock is dependent on a region at the C-terminus of the molecule since, when this region is deleted, HSF spontaneously trimerizes and can bind to DNA even in the absence of heat shock (Rabindran *et al.*, 1993). The C-terminal region contains a motif known as the leucine zipper, which contains a leucine residue every seven amino acids. As leucine zippers are known to be able to interact with

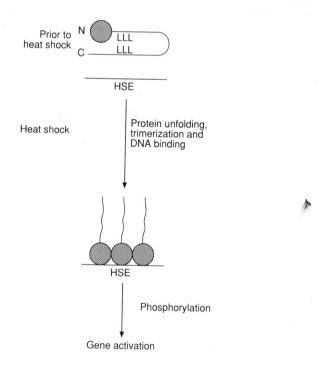

Figure 4.5 Prior to heat shock, HSF is present in a monomeric form in which the leucine zipper motifs (L) at the C-terminus and within the molecule promote intramolecular folding, which masks the N-terminal DNA-binding domain (shaded), preventing binding to the HSE. Following heat shock, the protein unfolds and forms the DNA-binding trimeric form. This form binds to the HSE and activates transcription following its subsequent phosphorylation.

one another (see Section 8.4), it is thought that this region acts by interaction with another leucine zipper located adjacent to the N-terminal DNA-binding domain promoting intra-molecular folding which masks the DNA binding domain. Following heat shock, HSF unfolds, unmasking the DNA-binding domain and allowing a DNA-binding trimer to form (Figure 4.5).

The two-stage process described above represents a common mechanism for the activation of HSF in higher eukaryotes such as *Drosophila* and mammals. In contrast, however, the *Saccharomyces cerevisiae* (budding yeast) HSF is activated by a much simpler mechanism. Thus, unlike *Drosophila* or mammalian HSF, the budding yeast protein lacks the C-terminal leucine zipper region which promotes monomer formation and therefore exists as a trimer prior to heat shock. As expected from this, HSF can be observed bound to the HSE even in non-heat-shocked cells (Sorger *et al.*, 1987). HSF can activate transcription, however, only after heat treatment, when the protein becomes phosphorylated. Interestingly, in *Schizosaccharomyces pombe* (fission yeast), HSF regulation follows the *Drosophila* and mammalian system, with HSF becoming bound to DNA only following heat shock (Gallo *et al.*, 1991).

Hence in mammals, *Drosophila* and fission yeast, activation of HSF is more complex than in budding yeast, involving an initial stage activating the DNA-binding ability of HSF in response to heat as well as the stage, common to all organisms, in which the ability to activate transcription is stimulated by phosphorylation (Figure 4.6). Such phosphorylation is suggested to result in the creation of a highly negatively

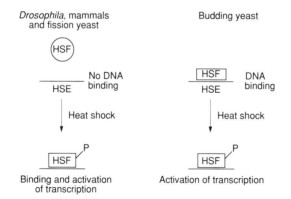

Figure 4.6 HSF activation in *Drosophila*, mammals and fission yeast compared with that in budding yeast. Note that in budding yeast HSF is already bound to DNA prior to heat shock and hence its activation by heat involves only the second of the two stages seen in other organisms, namely its phosphorylation, allowing it to activate transcription.

charged region on the HSF. This region would be analogous to the negatively charged acidic activation domains of other transcription factors (see Section 9.2.2). It would therefore allow HSF to interact with components of the constitutive transcriptional apparatus such as TFIID and activate transcription of the heat-shock genes. In agreement with this idea, HSF has been shown to possess a heat-inducible activation domain (Nieto-Sotelo *et al.*, 1990); for discussion see Parker (1989). Interestingly, this domain can be converted into a form that activates transcription at any temperature by deletion of a seven-amino-acid peptide which is located adjacent to the activation domain within HSF, indicating that this peptide normally inhibits the activity of the activation domain in a temperature-dependent manner (Jakobsen and Pelham, 1991).

In summary, therefore, heat shock induces the increased transcription of a small number of cellular genes via the post-translational modification of a pre-existing transcription factor, HSF. This factor binds to a sequence known as the HSE which is located in a region of the heat-shock gene promoters that is free of nucleosomes and contains a DNAse I-hypersensitive site even prior to heat shock. The activated form of HSF is capable, after binding, of interacting with components of the constitutive transcriptional apparatus which are bound at other sites in the promoter region, and thereby stimulating transcription.

4.3 STEROID-INDUCIBLE TRANSCRIPTION

4.3.1 Steroid receptors

The steroid hormones are a group of substances derived from cholesterol; they exert a wide range of effects on biological processes such as growth, metabolism and sexual differentiation (review: King and Mainwaring, 1974). Early studies using radioactively labelled hormones showed that they act by interacting with specific receptor proteins. This binding of hormone to its receptor activates the receptor and allows it to bind to a limited number of specific sites in chromatin. In turn this DNA binding activates transcription of genes carrying the receptor-binding site. Hence, as with the heat-shock factor, these receptor proteins are transcription factors becoming activated in response to a specific signal and in turn activating specific genes (reviews: Beato, 1989; Parker, 1993). These receptor proteins were therefore amongst the earliest transcription factors to be identified, well before the tech-

Table 4.2 Relationship of consensus sequences conferring responsivity to various hormones

Glucocorticoid/progesterone	GGTACANNNTGTTCT
Oestrogen	AGGTCANNNTGACCT
Thyroid hormone/retinoic acid	TCAGGTCA---TGACCTGA

N indicates that any base can be present at this position; a dash indicates that no base is present, the gap having been introduced to align the sequence with the other sequences.

niques described in Chapter 2 were in routine use, simply on the basis of their ability to bind radioactively labelled steroid ligand.

As with heat shock, genes that are induced by a particular steroid hormone contain a specific binding site for the receptor–hormone complex. The responses to different hormones such as glucocorticoids and oestrogen are mediated by distinct palindromic sequences which are related to one another. In turn, such sequences are related to the sequence that mediates induction by the related substances thyroid hormone and retinoic acid (Table 4.2) (review: Beato, 1989).

The basis of this binding site relationship was revealed when the genes encoding the receptor proteins were cloned. Thus they were found to constitute a family of genes encoding closely related proteins of similar structure, with particular regions being involved in DNA binding, hormone binding and transcriptional activation (Figure 4.7). This has led to the idea that these receptors are encoded by an evolutionarily related gene family which is known as the steroid–thyroid hormone receptor gene superfamily (reviews: Evans, 1988; Ham and Parker, 1989; Tsai and O'Malley, 1994).

As shown in Figure 4.7, the most conserved region between the different receptors is the DNA-binding domain explaining the ability of the receptors to bind to similar DNA sequences. Clearly, the differences in amino acid sequences between the DNA-binding motifs in the different receptors will determine the precise DNA sequence that each receptor recognizes. The replacement of particular amino acids in the DNA-binding domain with their equivalents in other receptors has therefore provided considerable information on the features that determine the precise sequence recognized or the optimal spacing between the two halves of the palindromic sequence. These experiments and the so-called 'zinc finger' motifs in the different receptors that mediate their DNA binding are discussed in Section 8.3.2. Interestingly, both DNAse I protection and methylation studies support the idea that the receptor binds to DNA as a dimer, each receptor molecule binding to one half of the palindromic recognition sequence.

Having established, therefore, that steroid hormones exert their effect on gene expression via specific receptor proteins, two questions remain. These are, firstly, how does binding of hormone to the receptor activate

Figure 4.7 Domain structure of individual members of the steroid–thyroid hormone receptor superfamily. The proteins are aligned on the DNA-binding domain, which shows the most conservation between different receptors. The percentage homologies in each domain of the receptors to that of the glucocorticoid receptor are indicated.

its ability to bind to specific DNA sequences and, secondly, how does such binding activate transcription? These questions will be considered in turn.

4.3.2 Activation of the receptor

Following identification of the hormone receptors, it was rapidly shown that the receptors were only found associated with DNA after hormone treatment. These early studies were subsequently confirmed by using DNAse I footprinting on whole chromatin to show that the receptor was bound to the hormone response sequence only following hormone treatment (Becker *et al.*, 1986). These studies were therefore consistent with a model in which the hormone induces a conformational change in the receptor, activating its ability to bind to DNA and thereby activate transcription.

Subsequent studies have suggested that the situation is more complex,

Figure 4.8 DNAse I footprint analysis of the binding of the glucocorticoid receptor to the glucocorticoid-inducible mouse mammary tumour virus long terminal repeat promoter (MMTV-LTR). In tracks I and II the DNAse I digestion has been carried out without any added receptor. In tracks 1–3, glucocorticoid receptor has been added prior to DNAse I digestion, either alone (track 1+), with the glucocorticoid hormone corticosterone (track 2) or with the anti-hormone RU486, which inhibits steroid-induced activation of the receptor (track 3). Track 1− shows the result of adding receptor to the DNA in the absence of DNAse I addition, in which some cleavage by endogenous nucleases (e) occurs, whilst track G is a marker track produced by cleaving the same DNA at each guanine residue. Minus signs indicate footprinted regions protected by receptor; plus signs are hypersensitive sites at which cleavage is increased by the presence of the receptor. The DNA fragment used and position of the radioactive label (diamond) are shown, together with the distances upstream from the initiation site for transcription. Note that the identical footprint is produced by the receptor either alone or in the presence of hormone or anti-hormone. Hence, *in vitro*, the receptor can bind to DNA in the absence of hormone.

Figure 4.9 Comparison of steroid receptor binding to DNA in the presence or absence of hormone *in vivo* and *in vitro*. Note that, whilst *in vivo* DNA binding can occur only in the presence of hormone, *in vitro* it can occur in the presence or absence of hormone.

however. Thus, although in the intact cell the receptor binds to DNA only in the presence of the hormone, purified receptor can bind to DNA *in vitro* in a band-shift or footprinting assay regardless of whether hormone is present or not (Wilmann and Beato 1986; Figures 4.8 and 4.9).

This discrepancy led to the suggestion that the receptor is inherently capable of binding to DNA but is prevented from doing so in the absence of steroid because it is anchored to another protein. The hormone acts to release it from this association and allow it to fulfil its inherent ability to bind to DNA. In agreement with this possibility, in the absence of hormone the glucocorticoid receptor protein is found in the cytoplasm complexed to a 90 000 molecular weight heat-inducible protein (hsp90) in an 8S complex. This complex is dissociated upon steroid treatment, releasing the 4S receptor protein (reviews: Pratt, 1993; Smith and Toft, 1993). The released receptor is free to dimerize and move into the nucleus. Since these processes have been shown to be essential for DNA binding and transcriptional activation by steroid hormone receptors (Tsai *et al.*, 1988; Fawell *et al.*, 1990), dissociation of the receptor from hsp90 is essential if gene activation is to occur. In agreement with this, antiglucocorticoids that inhibit the positive action of glucocorticoids have been shown to stabilize the 8S complex of hsp90 and the receptor (Groyer *et al.*, 1987). Similar complexes with hsp90 have also been reported for the other steroid hormone receptors. Thus the activation of the different steroid receptors by their specific hormones is likely to involve disruption of the protein–protein interaction with hsp90 (Figure 4.10). The manner in which the steroid achieves this effect is discussed further in Section 10.3.3.

In addition to the steroid-induced dissociation of the receptors from

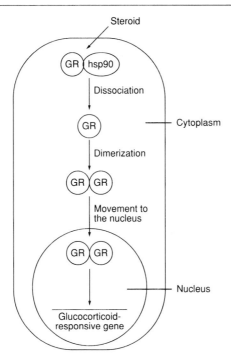

Figure 4.10 Activation of the glucocorticoid receptor (GR) by steroid involves dissociation of hsp90, allowing dimerization and movement to the nucleus.

hsp90, recent *in vitro* studies have suggested that a second step following dissociation from hsp90 may also be required for receptor activation. Thus, in a cell free system in which the progesterone receptor exists in a 4S form, free of bound hsp90, the addition of progesterone is still required for the activation of progesterone responsive genes (Bagchi *et al.*, 1990). This indicates that the hormone has an additional effect on the receptor apart from dissociating it from hsp90. This effect is likely to involve the unmasking of a previously inactive transcriptional activation domain in the receptor, allowing it to activate gene expression in a hormone-dependent manner following DNA binding. Thus, domain-swapping experiments (see Section 2.3.3) have identified regions in both the glucocorticoid and oestrogen receptors which, when linked to the DNA-binding domain of another factor, can activate transcription only after hormone addition. These regions hence constitute hormone-dependent activation domains (Webster *et al.*, 1988). Moreover, in the case of the oestrogen receptor, it has been shown that the oestrogen antagonist 4-hydroxytamoxifen induces the receptor to bind to DNA (presumably by promoting dissociation from hsp90 and dimerization) but does not induce gene activation (Kumar and Chambon, 1988),

Figure 4.11 Activation of the steroid receptors by treatment with steroid. As well as inducing dissociation of the receptor from hsp90, steroid treatment also increases the ability of the receptor to activate transcription following DNA binding.

suggesting that it fails to activate the oestrogen-responsive trans-activation domain. Hence the mechanism by which the receptors are activated is now thought to involve both dissociation from hsp90 and a change in their transcriptional activation ability (Figure 4.11). The various means by which steroid mediates receptor activation are discussed further in Section 10.3.3.

4.3.3 Activation of transcription

Having considered the activation of the receptor by hormone, it is necessary to discuss how binding of the activated receptor in turn activates transcription of the genes bearing its target site (review: Tsai and O'Malley, 1994).

As discussed in Section 4.2.2 the heat-shock factor binds to a region of chromatin that is free of histone-containing nucleosomes and is hypersensitive to DNAse I digestion. This hypersensitive site exists prior to heat shock, marking the potential binding site for the transcription factor. In contrast, in a number of cases, steroid hormone treatment has been shown to cause the induction of a DNAse I-hypersensitive site located at the DNA sequence to which the receptor binds. Hence the binding of the receptor may activate transcription by displacing or altering the structure of a nucleosome within the promoter of the gene, creating the hypersensitive site. In turn this would facilitate the binding of other transcription factors necessary for gene activation whose binding sites would be exposed by the change in the position or structure of the nucleosome. These factors would be present in the cell in an active form prior to steroid treatment but could not bind to the gene because their binding sites were masked by a nucleosome (Figure

Steroid-inducible genes

Uninduced

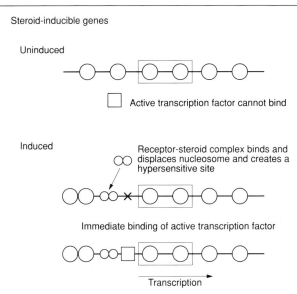

☐ Active transcription factor cannot bind

Induced

Receptor-steroid complex binds and
displaces nucleosome and creates a
hypersensitive site

Immediate binding of active transcription factor

Transcription

Figure 4.12 Binding of the steroid receptor–steroid hormone complex to the promoter of a steroid-inducible gene results in a change in chromatin structure, creating a hypersensitive site and allowing pre-existing constitutively expressed transcription factors to bind and activate transcription. Note that the binding of these constitutive factors may occur because the receptor totally displaces a nucleosome from the DNA, as shown in the diagram, or because it alters the structure of the nucleosome so as to expose specific binding sites in the DNA.

4.12) (for discussion see Travers, 1994). In agreement with this idea, the binding sites for TFIID and CTF/NF1 (see Sections 3.2.3 and 3.3.3) in the glucocorticoid-responsive mouse mammary tumour virus promoter are occupied only after hormone treatment, although these factors are present in an active DNA-binding form at a similar level in treated and untreated cells.

This mechanism, in which the receptor acts by displacing or altering the structure of a nucleosome allowing constitutive factors access to their binding sites, is clearly in contrast to the binding of HSF to a promoter which already lacks a nucleosome and contains bound GAGA factor and TFIID. In this latter case, activation of transcription must occur not via alteration in chromatin structure but via interaction with the components of the constitutive transcriptional apparatus. It should be noted, however, that these two mechanisms are not exclusive. Thus, as discussed above (Section 4.3.2), the steroid receptors contain specific activation domains which can activate transcription when linked to the DNA-binding regions of other factors. As discussed in Section 1.3, such activation domains are known to act by interacting with other bound transcription factors to increase transcription. This

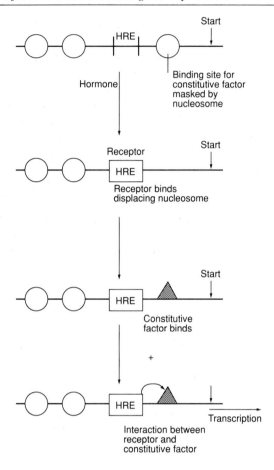

Figure 4.13 Activation of steroid-inducible genes by steroid receptors. As well as altering chromatin structure and allowing constitutive factors to bind, the hormone receptor is also able to interact directly with constitutive factors and increase their ability to activate transcription.

finding indicates, therefore, that the steroid receptors promote transcription both by altering chromatin structure to allow constitutive factors to bind and also by interacting directly with these other factors (Figure 4.13). The activation of transcription by factors that alter chromatin structure and/or interact with the basal transcriptional complex is discussed further in Sections 9.2.3 and 9.2.4.

Hence, in the case of the steroid receptors, a single factor can alter the chromatin structure and then activate transcription. In contrast in heat-shock gene activation, these functions are performed by separate factors, with the GAGA factor displacing a nucleosome, allowing HSE to bind and activate transcription following a subsequent heat shock.

4.3.4 Inhibition of gene expression by steroid hormones

In the preceding sections we have discussed gene activation mediated by steroid hormones complexed to their corresponding receptors. In addition, however, these hormones can also inhibit the expression of specific genes, glucocorticoid for example, inhibiting expression of the genes encoding bovine prolactin and human pro-opiomelanocortin (POMC). The inhibitory effect observed in these cases is mediated by binding to DNA of the identical receptor–hormone complex which activates glucocorticoid-inducible genes. However, the DNA sequence element to which the complex binds when mediating its negative effect (nGRE) is distinct from the glucocorticoid response element (GRE) to which it binds when inducing gene expression, although the two are related (Figure 4.14).

This has led to the suggestion that the sequence difference causes the receptor–hormone complex to bind to the nGRE in a configuration in which its activation domain cannot interact with other transcription factors to activate transcription, as occurs following binding to the positive element (Sakai *et al.*, 1988; Figure 4.15). In agreement with this idea, the glucocorticoid receptor has been shown to bind to the

Binding site for positive regulation G G T A C A N N N T G T T C T

Binding site for negative regulation A T Y A C N N N N T G A T C W

Figure 4.14 Relationship of the sites in DNA that mediate gene activation or repression by binding the glucocorticoid receptor. Note that the sites are related but distinct.

Figure 4.15 Consequences of glucocorticoid receptor binding to the DNA-binding sites that mediate gene activation (GRE) or repression (nGRE). Note that the receptor is likely to bind in a different configuration to the two different sequences, resulting in its ability to activate transcription only following binding to the GRE.

nGRE in the POMC gene as a trimer rather than the dimer form which binds to the GRE and stimulates transcription (Drouin *et al.*, 1993). The receptor bound in this configuration to the negative element may exert a direct negative influence resulting in decreased transcription of the gene. More probably, however, it simply acts by preventing binding of a positive acting factor to this or an adjacent site, thereby preventing gene induction. In agreement with this idea, the nGRE in the human glycoprotein hormone α-subunit gene which overlaps a cyclic AMP response element (CRE) is able to inhibit gene expression only when the CRE is left intact. Hence it is likely that receptor bound at the negative element prevents binding of a transcriptional activator to the CRE and thereby inhibits gene expression (Akerblum *et al.*, 1988; Figure 4.16).

In the case of the glucocorticoid receptor, inhibition of gene expression by binding to the nGRE occurs only after glucocorticoid treatment, which allows it to dissociate from hsp90 and bind to DNA. However, the thyroid hormone receptor, which does not complex to hsp90, can bind to its response element (TRE) in the absence of thyroid hormone and inhibit gene expression. This effect is not due to the receptor preventing other positive factors from binding but actually involves a direct inhibitory effect of the receptor on transcription, which requires a specific inhibitory domain at the C-terminus of the molecule (Baniahmad *et al.*, 1992). In the presence of thyroid hormone, the receptor undergoes a conformational change which exposes its activation domain and converts it from a repressor to an activator (Figure 4.17). Hence, in this case, gene activation or repression can be mediated from

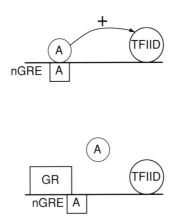

Figure 4.16 Inhibition of gene expression by glucocorticoid receptor binding to an nGRE is likely to be mediated by preventing the binding of a positively acting activator protein (A) to a site adjacent to or overlapping the nGRE.

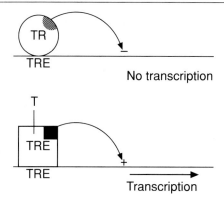

Figure 4.17 In the absence of thyroid hormone (T) the thyroid hormone receptor inhibits gene expression via a discrete inhibitory domain (hatched box). Binding of thyroid hormone (T) exposes the activation domain of the receptor (solid box) and allows it to activate transcription.

the same DNA-binding site, with the effect depending on the presence or absence of the hormone.

Interestingly, the thyroid hormone receptor can also inhibit gene activation by other members of the steroid–thyroid hormone receptor family. Thus, as shown in Table 4.2, the binding sites mediating responsiveness to oestrogen (ERE) and thyroid hormone (TRE) are identical in sequence and differ only in the spacing between the two halves of the palindrome which are adjacent in the TRE and separated by three bases in the ERE. Because of this sequence similarity, the thyroid hormone receptor can bind to an ERE but, presumably due to a difference in its

Figure 4.18 The thyroid hormone receptor can bind to a site mediating gene activation by the oestrogen receptor (ERE) but cannot activate transcription. As this binding inhibits binding of the oestrogen receptor itself to the ERE, it results in the inhibition of oestrogen receptor-mediated gene activation.

configuration when bound to a gapped site, it cannot activate transcription even in the presence of thyroid hormone. It does, however, inhibit gene activation by oestrogen since it prevents binding of the oestrogen receptor–hormone complex to the ERE (Glass *et al.*, 1988; Figure 4.18). Similarly the identical structure and sequence of the sites mediating thyroid hormone and retinoic acid responsiveness (Table 4.2) allows the thyroid hormone receptor to bind to a TRE in the absence of thyroid hormone and thereby inhibit gene induction by retinoic acid and its receptor both by direct inhibition and by preventing binding of the activated retinoic acid receptor (Graupner *et al.*, 1989; Figure 4.19).

These examples illustrate, therefore, how interacting networks of transcription factors, each capable of being activated by hormones or other signals, can produce complex patterns of gene activation and repression. Interestingly, however, these interactions can also take place between different forms of the same hormone receptor. Thus,

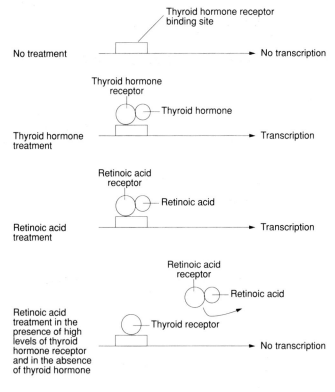

Figure 4.19 The thyroid hormone receptor can bind to its DNA-binding site in the absence of thyroid hormone and inhibit transcription directly, as well as preventing binding of the retinoic acid–retinoic acid receptor complex to the identical binding site, thereby preventing gene activation in response to retinoic acid.

the thyroid hormone receptor exists in two alternatively spliced forms. These two forms differ in their ability to bind thyroid hormone with one form (α2) lacking a part of the hormone-binding domain and therefore unable to bind hormone (Koenig *et al.*, 1989; Figure 4.20a). Both the α2 form and the hormone-binding α1 form can bind to DNA, however. Binding of α2 to the TRE sequence prevents binding of α1 and thereby prevents gene induction in response to thyroid hormone (Figure 4.20b). As discussed in Section 7.2.2, a similar non-hormone-binding form of the thyroid hormone receptor is encoded by the v-*erbA* oncogene, which produces cancer by inhibiting the expression of thyroid hormone-responsive genes involved in erythroid differentiation.

It should be noted that all the examples of repression by members of the steroid–thyroid receptor family discussed so far involve their binding to specific sequences in the DNA, such as the nGRE or the specific

Figure 4.20 (a) Relationship of the ErbA α1 and α2 proteins. Note that only the α1 protein has a functional thyroid hormone-binding domain. (b) Inhibition of ErbA α1 binding and of gene activation in the presence of the α2 protein.

binding sites for other members of the family. It is also possible, however, for receptors such as the glucocorticoid receptor to produce repression without binding to DNA by forming a protein–protein complex with another transcription factor. Thus glucocorticoid hormones have been known for some time to be potent inhibitors of the induction of the collagenase gene by phorbol esters, resulting in their having an anti-inflammatory effect. This inhibition is mediated by the glucocorticoid receptor, which abolishes the binding of the Jun and Fos proteins to the AP1 sites in the collagenase gene. Since such binding is essential for gene induction by phorbol esters (for discussion of Fos, Jun and AP1 see Section 7.2.1), this effectively inhibits gene activation. Unlike the examples of repression by the glucocorticoid receptor discussed above, however, the collagenase promoter does not contain any binding sites for the receptor adjacent to the AP1 sites, nor does the receptor apparently bind to the collagenase promoter (Jonat *et al.*, 1990). Rather, repression is mediated via the formation of a protein–protein complex between Fos or Jun and the glucocorticoid receptor, which prevents binding of Fos and Jun to the AP1 site and hence inhibits transcription (Figure 4.21). Clearly such an association of Fos or Jun and the glucocorticoid receptor will also prevent the glucocorticoid receptor from binding to DNA

Figure 4.21 Mutual transrepression by Fos/Jun and the glucocorticoid receptor. The formation of a protein–protein complex between Fos/Jun and the glucocorticoid receptor inhibits the expression of genes containing binding sites for either Fos/Jun (AP1 sites) or for the glucocorticoid receptor (GRE) by preventing the proteins binding to these sites and activating transcription.

and thereby activating glucocorticoid-dependent genes in response to hormone; this is indeed observed at high concentrations of Fos and Jun (Jonat *et al.*, 1990).

Hence mutual transrepression of two different activating proteins can be achieved by their forming a complex that inhibits the DNA binding of both. Moreover, this mutual repression illustrates how different cellular signalling pathways, which are activated respectively by phorbol esters and glucocortocoid hormones, can interact with one another, resulting in cross-talk between the pathways (review: Schule and Evans, 1991).

So, as well as being able to activate gene expression, the members of the steriod–thyroid hormone receptor family also illustrate three mechanisms by which repression of gene expression can be achieved, namely the direct inhibition of transcription or the neutralization of a positive factor by preventing its binding to DNA either by masking of its site or by interacting with it to form a non-DNA-binding complex (for further discussion of the mechanisms of transcriptional repression see Section 9.3).

4.4 CONCLUSIONS

This chapter has discussed in detail two systems in which specific genes are switched on in response to a particular inducer. As indicated in Table 4.1, many other inducing agents can activate particular genes containing specific DNA sequences. The examples described here, however, illustrate the general principles by which such induction takes place.

Thus the induction event normally involves the activation of a particular transcription factor which is present in the cell in an inactive form prior to induction. This activation of the transcription factor may take place via post-translational modifications such as phosphorylation, a change in the conformation of the factor and/or disruption of an inhibitory protein–protein interaction.

In turn, such modifications may produce activation of the factor by allowing it to move to the nucleus and fulfil its inherent DNA-binding ability, as in the case of the steroid receptors. Alternatively they may convert the protein from a form that cannot bind DNA into a DNA-binding form, as in the case of the *Drosophila*, mammalian and fission yeast heat-shock factors. Lastly the conversion to an active form may not involve any alteration in DNA binding at all but may enhance the ability of bound factor to enhance transcription, as in the case of the budding yeast HSF. These activating effects of inducing agents on transcription factors will be discussed further in Chapter 10 as part of a general discussion on the activation of transcription factors in both tissue-specific and inducible gene expression, and in development.

Once the transcription factor has been activated, it induces gene expression by interacting with the constitutive components of the transcriptional apparatus discussed in Chapter 3 to enhance the stability of the stable transcription complex and hence increase transcription. This process, which is characteristic of virtually all factors mediating inducible gene expression, is observed for both the heat-shock factor and the steroid receptors, which contain activation domains capable of interacting with other transcription factors. In addition, however, the steroid receptors illustrate another facet of gene activation not seen in the heat-shock factor. Thus the binding of the receptor to DNA results in an alteration of chromatin structure. This exposes the DNA-binding sites for constitutive factors such as CTF/NF1, allowing these factors to bind to their exposed binding sites in DNA and activate transcription. In the case of the heat-shock genes, this opening of the chromatin has already occurred due to the binding of the GAGA factor, allowing HSE to bind and activate transcription following heat shock.

The activation of the various transcription factors in response to the inducing stimulus, and the consequent effect of the activated factors on the constitutive transcriptional apparatus, therefore results in the observed inducible pattern of gene expression.

REFERENCES

Akerblum, I. W., Slater, E. P., Beato, M., Baxter, J. D. and Mellon, P. L. (1988). Negative regulation by glucocorticoids through interference with a cAMP responsive enhancer. *Science* **241**, 350–353.

Bagchi, M. K., Tsai, S. Y., Tsai, M.-J. and O'Malley, B. W. (1990). Identification of a functional intermediate in receptor activation in progesterone dependent cell-free transcription. *Nature* **345**, 547–550.

Baniahmad, A., Kohne, A. C. and Renkawitz, R. (1992). A transferable silencing domain is present in the thyroid hormone receptor, in the v-*erbA* oncogene product and in the retinoic acid receptors. *EMBO Journal* **11**, 1015–1023.

Beato, M. (1989). Gene regulation by steroid hormones. *Cell* **56**, 335–344.

Becker, P. B., Gloss, B., Schmid, W., Strahle, U. and Schutz, G. (1986).

In vivo protein–DNA interactions on a glucocorticoid response element require the presence of the hormone. *Nature* **324**, 686–688.

Drouin, J., Sun, Y. L., Chamberland, M., Ganthier, Y., De Lean, A., Nemer, M. and Schmidt, T. J. (1993). Novel glucocorticoid receptor complex with DNA element of the hormone repressed POMC gene. *EMBO Journal* **12**, 145–156.

Evans, R. M. (1988). The steroid and thyroid hormone receptor gene super-family. *Science* **240**, 889–895.

Fawell, S. E., Lees, J. A., White, R. and Parker, M. G. (1990). Characterization and colocalization of steroid binding and dimerization activities in the mouse oestrogen receptor. *Cell* **60**, 953–962.

Gallo, G. J., Schuetz, T. J. and Kingston, R. E. (1991). Regulation of heat shock factor in *Schizosaccharomyces pombe* more closely resembles regulation in mammals than in *Saccharomyces cerevisiae*. *Molecular and Cellular Biology* **11**, 281–288.

Glass, C. K., Holloway, I. M., Devary, O. V. and Rosenfeld, M. G. (1988). The thyroid hormone receptor binds with opposite transcriptional effect to a common sequence motif in thyroid hormone and estrogen response elements. *Cell* **54**, 313–323.

Graupner, G., Wills, K. N., Tzukerman, M., Zhang, X.-K. and Pfahl, M. (1989). Dual regulatory role for thyroid-hormone receptors allows control of retinoic acid receptor activity. *Nature* **340**, 653–656.

Groyer, A., Schweizer-Groyer, G., Cadepond, F., Mariller, M. and Baullieu, E. E. (1987). Antiglucocorticosteroid effects suggest why steroid hormone is required for receptors to bind DNA *in vivo* but not *in vitro*. *Nature* **328**, 624–626.

Ham, J. and Parker, M. G. (1989). Regulation of gene expression by hormone receptors. *Current Opinion in Cell Biology* **1**, 503–511.

Hensold, J. O., Hunt, C. R., Calderwood, S. K., Housman, D. E. and Kingston, R. E. (1990). DNA binding of heat shock factor to the heat shock element is insufficient for transcriptional activation in murine erythroleukemia cells. *Molecular and Cellular Biology* **10**, 1600–1608.

Jakobsen, B. K. and Pelham, H. R. B. (1991). A conserved heptapeptide restrains the activity of the yeast heat shock transcription factor. *EMBO Journal* **10**, 369–375.

Jonat, C., Rahmsdorf, H. G., Park, K. K., Cato, A. C. B., Gebel, S., Punta, H., and Herrlich, P. (1990). Antitumour promotion and anti inflammation: down modulation of AP–1 (FOS/JUN) activity by glucocorticoid hormone. *Cell* **62**, 1189–1201.

Jones, N. C., Rigby, P. W. J. and Ziff, E. B. (1988). Trans-acting protein factors and the regulation of eukaryotic transcription: lessons from studies on DNA tumour viruses. *Genes and Development* **2**, 267–281.

King, R. J. B. and Mainwaring, W. I. P. (1974). *Steroid Cell Interactions*. London: Butterworths.

Koenig, R. G., Lazar, M. A., Hodin, R. A., Brent, G. A., Larsen, P. R., Chin, W. W. and Moore, D. D. (1989). Inhibition of thyroid hormone action by a non-hormone binding *c-erbA* protein generated by alternative RNA splicing. *Nature* **337**, 659–661.

Kumar, V. and Chambon, P. (1988). The estrogen receptor binds tightly to its responsive element as a ligand-induced homodimer. *Cell* **55**, 145–156.

Larson, J. S., Schuetz, T. J. and Kingston, R. E. (1988). Activation *in vitro* of sequence-specific DNA binding by a human regulatory factor. *Nature* **335**, 372–375.

Latchman, D. S. (1995). *Gene Regulation: A Eukaryotic Perspective*, 2nd ed. London: Chapman and Hall.

Maniatis, T., Goodburn, S. and Fischer, J. A. (1987). Regulation of inducible and tissue-specific gene expression. *Science* **236**, 1237–1245.

Morimoto, R. (1993). Cells in stress: transcriptional activation of heat shock genes. *Science* **259**, 1409–1410.

Nieto-Sotelo, J., Wiederrecht, G., Okuda, A. and Parker, C. S. (1990). The yeast heat shock transcription factor contains a transcriptional activation domain whose activity is repressed under non shock conditions. *Cell* **62**, 807–817.

Parker, C. S. (1989). Transcription factors. *Current Opinion in Cell Biology* **1**, 512–518.

Parker, M. G. (1993). Steroid and related receptors. *Current Opinion in Cell Biology* **5**, 499–504.

Parsell, D. A. and Lindquist, S. (1993). The function of heat shock proteins in stress tolerance: degradation and reactivation of damaged proteins. *Annual Review of Genetics* **27**, 437–496.

Pratt, W. B. (1993). The role of heat shock proteins in regulating the function, folding and trafficking of the glucocorticoid receptor. *Journal of Biological Chemistry* **269**, 21455–21458.

Rabindran, S. K., Haroun, R. I., Clos, J., Wisniewski, J. and Wu, C. (1993). Regulation of heat shock factor trimer formation: role of a conserved Leucine zipper. *Science* **259**, 230–234.

Sakai, D. D., Helms, S., Carlstedt-Duke, J., Gustafsson, J.-A., Rottman, F. M. and Yamamoto, K. R. (1988). Hormone-mediated repression: a negative glucocorticoid response element from the bovine prolactin gene. *Genes and Development* **2**, 1144–1154.

Schule, R. and Evans, R. M. (1991). Cross coupling of the signal transduction pathways. Zinc finger meets Leucine zipper. *Trends in Genetics* **7**, 377–381.

Smith, D. F. and Toft, D. O. (1993). Steroid receptors and their associated proteins. *Molecular Endocrinology* **7**, 8–11.

Sorger, P. K., Lewis, M. J. and Pelham, H. R. B. (1987). Heat shock factor is regulated differently in yeast and HeLa cell. *Nature* **329**, 81–84.

Travers, A. A. (1994). Chromatin structure and dynamics. *Bioessays* **16**, 657–662.

Tsai, M.-J. and O'Malley, B. W. (1994). Molecular mechanisms of action of steroid/thyroid receptor super family members. *Annual Review of Biochemistry* **63**, 451–486.

Tsai, S. Y., Carlstedt-Duke, J., Weigel, N. L., Dalman, K., Gustafsson, J. A., Tsai, M.-J. and O'Malley, B. W. (1988). Molecular interactions of steroid hormone receptor with its enhancer element: evidence for receptor dimer formation. *Cell* **55**, 361–369.

Tsukiyama, T., Becker, P. B. and Wu, C. (1994). ATP-dependent nucleosome disruption at a heat-shock promotor mediated by DNA binding of GAGA transcription factor. *Nature* **367**, 525–532.

Van Holde, K. (1994). Gene expression: properly preparing promotors. *Nature* **367**, 512–513.

Webster, N. J. G., Green, S., Jin, J. R. and Chambon, P. (1988). The hormone-binding domains of the estrogen and glucocorticoid receptors contain an inducible transcription activation function. *Cell* **54**, 199–207.

Wilmann, T. and Beato, M. (1986). Steroid-free glucocorticoid receptor binds specifically to mouse mammary tumour DNA. *Nature* **324**, 688–691.

Wu, C. (1985). An exonuclease protection assay reveals heat shock element and TATA-box binding proteins in crude nuclear extracts. *Nature* **317**, 84–87.

Zimarino, V. and Wu, C. (1987). Induction of sequence-specific binding of *Drosophila* heat-shock activator protein without protein synthesis. *Nature* **327**, 727–730.

CHAPTER FIVE

Transcription factors and cell type-specific transcription

5.1 CELL TYPE-SPECIFIC GENE EXPRESSION

As discussed in Chapter 4, both prokaryotes and eukaryotes respond to specific stimuli by inducing the expression of particular genes, this process being mediated by transcription factors which are activated in response to the stimulus. In eukaryotes, however, transcription factors also play a critical role in processes that have no parallel in prokaryotes. Thus, the higher eukaryote contains a vast range of different cell types, each of which expresses specific genes encoding particular products necessary for the specialized function of that cell type. The role of transcription factors in controlling the cell type-specific expression of particular genes is the subject of this chapter.

A number of different factors involved in mediating gene expression in specific cell types have been defined; a selection of these is listed in Table 5.1. These factors are normally synthesized or activated only in one specific tissue, resulting in the cell type-specific transcription of genes whose expression is dependent upon them.

In general such factors have been defined by studying the regulation of a specific, well characterized, gene which is expressed only in a particular cell type and by identifying the cell type-specific transcription factor responsible for this pattern of expression. Subsequently, once this factor has been identified, its role in the regulation of other, less well characterized, genes expressed in the particular cell type can be assessed. For example, the transcription factor GATA-1 (NF-E1), which is expressed at very high levels in erythroid cells, was originally identi-

Table 5.1 Examples of transcription factors regulating cell type-specific gene expression in mammals

Factor	Cell type	Gene regulated
EF1	Erythroid	Haemoglobin, porphobilinogen deaminase
Isl-1	Islet cells of the pancreas	Insulin
LFB1	Liver	Albumin, α_1-antitrypsin, fibrinogen
MyoD	Skeletal muscle	Myosin, creatine kinase
NFκB	B cells, activated T cells	Immunoglobin κ light chain, IL-2 α-receptor
NFATI	Activated T cells	Interleukin 2
Oct-2	B cells	Immunoglobin heavy and light chains
Pit-1	Anterior pituitary	Growth hormone

fied on the basis of its binding to the promoter or enhancer regions of chicken and mammalian globin genes, and was shown to be essential for their erythroid-specific pattern of gene expression, deletion or mutation of EF1-binding sites resulting in the abolition of such expression (Reitman and Felsenfeld, 1988). Subsequently this protein has also been implicated in the erythroid-specific gene expression of the porphobillinogen deaminase gene, which also contains a binding site for the factor within its promoter (Mignotte *et al.*, 1989). Finally, it was shown that the introduction of a mutation into the GATA-1 gene, which inactivated it, led to a failure of red blood cell formation, although all other cell types, including white blood cells, were still produced (Pevny *et al*, 1991). This factor therefore plays a critical role in the regulation of a number of different erythroid-specific genes and thereby in erythroid cell differentiation.

In Section 5.2 we will consider one example of the role of transcription factors in tissue-specific gene expression which has been defined in this way. This involves the B cell-specific expression of the genes encoding the heavy and light chains of the immunoglobulin molecule which is regulated by two transcription factors that are specifically active in B cells, namely NFκB and Oct-2.

However, not all tissue-specific transcription factors have been identified in this manner. Thus, the factor MyoD, which activates muscle-specific genes, was not identified on the basis of studying genes of this type. Rather, it was isolated on the basis of the finding that its artificial expression within undifferentiated fibroblast-like cells was sufficient to transform them into muscle precursor cells (Davis *et al.*, 1987). Hence this factor is not only capable of switching on muscle specific genes but by doing so actually plays a central role in the production of the differentiated cell type itself. This role of MyoD in determining the muscle cell type is discussed in Section 5.3.

The role of transcription factors in producing specific cell types has been particularly well characterized in homothallic yeasts. In this case, whether a cell is a or α in mating type is determined by which transcription factor gene is expressed at the mating type locus. This system, which is discussed in Section 5.4, not only offers insights into cell type-specific gene expression but also provides a possible model system for studies on the developmental regulation of gene expression by transcription factors which will be discussed in Chapter 6.

5.2 TRANSCRIPTION FACTORS AND IMMUNOGLOBULIN GENE EXPRESSION

5.2.1 B cell-specific expression of the immunoglobulin genes

The B lymphocytes which circulate in the blood of higher vertebrates have as their primary role the production of antibodies to defend the body against foreign organisms etc. These antibodies are produced by the association of two immunoglobulin heavy chains and two immunoglobulin light chains to produce a functional antibody molecule (for a recent review of the immune system, emphasizing molecular aspects, see Roitt *et al.* (1993)). Consistent with the need to produce antibodies only in B cells, the separate genes encoding the heavy and light chains are expressed in a B cell-specific manner.

This B cell specificity is produced by the combined action of a promoter element located upstream of the transcribed region and an enhancer element located within the intervening sequence (intron) separating the exons encoding the variable and constant regions of the molecule (Figure 5.1). Both these elements, if tested in isolation by linkage to a marker gene, are capable of driving high-level expression in B cells and hence are both B cell specific. In combination, however, the promoter and the enhancer that acts on it constitute a powerful B cell-specific-regulatory element and produce the high-level expression of the immunoglobulin genes observed in B cells (Garcia *et al.*, 1986).

Interestingly, the promoter and enhancer are separated by a large distance in gene-line DNA and are brought together by a DNA rearrangement event which allows the enhancer to act on the promoter

Figure 5.1 Structure of the rearranged immunoglobulin genes in a B cell. The B cell-specific promoter (P) is located adjacent to the DNA segments encoding the variable (V) and joining (J) regions of the molecule. These DNA segments are separated from the DNA segment encoding the constant (C) region of the immunoglobulin molecule by an intervening sequence (intron) which contains the B cell-specific enhancer (E).

Figure 5.2 The rearrangement of the heavy-chain gene brings an enhancer (E) in the intervening sequence between J and C close to the promoter (P) adjacent to V, and results in activation of the promoter and gene transcription.

producing maximal gene expression (Figure 5.2). The role of this rearrangement in gene regulation is entirely secondary, however, to its primary role in producing antibody diversity by the combination of different DNA sequences encoding different variable, joining and constant regions to produce different antibody molecules (review: Gellert, 1992).

  Hence the B cell-specific expression of the immunoglobulin genes is controlled by the inherent B cell specificity of both their promoters and enhancers. An understanding of this B cell specificity requires an understanding, therefore, of the transcription factors binding to these promoters and enhancers, and of their role in gene expression.

5.2.2 Structure and activity of immunoglobulin promoter and enhancer elements

The promoters of the immunoglobulin heavy- and light-chain genes have a relatively simple structure containing only one identifiable

Figure 5.3 The immunoglobulin heavy- and light-chain gene promoter and adjacent transcribed region. The promoter contains a TATA box and an octamer motif. Note the opposite orientation of the octamer motif in the heavy- and light-chain genes relative to the start site of transcription. The initiation site of transcription, the start site (ATG) of the translated region (shaded) and the splice site at the end of the first exon are illustrated. Numbers indicate the average distance (in base pairs) between the different elements in the different immunoglobulin genes.

transcription factor-binding site in addition to the TATA box (Parslow *et al.*, 1984; Figure 5.3). This octamer motif, which has the consensus sequence ATGCAAAT, is found approximately 70 bases upstream of the transcription initiation site in both light- and heavy-chain genes. Interestingly, the octamer is oriented in the opposite direction relative to the start site of transcription in the heavy- and light-chain genes.

In both these genes the octamer element plays a critical role in the B cell-specific activity of the promoter. Thus its deletion or mutation, leaving the rest of the promoter intact, abolishes B cell-specific expression. Similarly linkage of the octamer motif to a non-immunoglobulin promoter results in a B cell-specific expression pattern of the heterologous promoter (Wirth *et al.*, 1987). Interestingly, the octamer motif is also found within the enhancer elements of the immunoglobulin genes. In this situation, also, deletion of the octamer diminishes the activity of the enhancer in B cells (Gerster *et al.*, 1987). Hence the B cell specificity of both immunoglobulin promoters and enhancers involves the octamer motif.

In DNA mobility shift assays the octamer motif binds two major proteins in B cell extracts (see Figure 2.2) which are known as Oct-1 and Oct-2 (Singh *et al.*, 1986). One of these proteins, Oct-1, is present in virtually all cell types, whereas the smaller Oct-2 protein is found in B cells that express the immunoglobulin genes and not in most other cell types (Figure 2.2).

This expression pattern immediately suggests that it is the binding of Oct-2 to the octamer motif in the immunoglobulin genes which determines their B cell-specific expression. In agreement with this idea, the level of gene expression directed by an octamer-containing enhancer element was severely reduced in B cells derived from mice that lack any expression of Oct-2. Moreover, the activity of this enhancer can be stimulated in these cells by introducing a DNA construct expressing Oct-2, whereas the enhancer was not stimulated by Oct-1 (Pfisterer *et al.*, 1994). Hence Oct-2, but not Oct-1, can directly stimulate an octamer-containing enhancer in B cells.

The specific expression of Oct-2 in B cells therefore plays a role in the B cell specificity of immunoglobulin gene expression. This tissue-specific expression of Oct-2 is paralleled by the presence of its corresponding mRNA at high levels in B cells whilst it is absent in most other cell types (Clerc *et al.*, 1988). Hence, unlike the heat-shock factor or the steroid receptors, Oct-2 is not activated from a pre-existing inactive form when it is required; rather, the Oct-2 mRNA and protein are synthesized in B cells but not in most other cell types. Oct-2-dependent expression of the immunoglobulin enhancer is therefore controlled by the tissue-specific synthesis of Oct-2 rather than by its tissue-specific activation.

Interestingly, however, in contrast to the immunoglobulin enhancer, the activity of the immunoglobulin promoter is not diminished in B cells from mice that lack Oct-2 (Pfisterer *et al.*, 1994), even though, as noted above, the deletion of the octamer motif from the promoter abolishes its B cell-specific activity. Hence, unlike the octamer motif in the enhancer, that in the promoter can be activated by Oct-1, although this occurs only in B cells even though Oct-1 is present in all cell types. The solution to this paradox is provided by the recent finding that a second B cell-specific factor which does not itself bind to the octamer motif can interact with DNA-bound Oct-1 or Oct-2 and activate transcription from an octamer-containing promoter (Gstaiger *et al.*, 1995; Strubin *et al.*, 1995). Hence the B cell-specific activity of the immunoglobulin genes is produced by the tissue-specific expression of this co-activator, which can interact with either Oct-2 or the constitutively expressed Oct-1 to activate transcription from the immunoglobulin promoter, as well as by the tissue-specific expression of Oct-2 itself which, unlike Oct-1, can activate transcription from the immunoglobulin enhancer.

Although the immunoglobulin promoter contains only an octamer motif, the immunoglobulin heavy-chain enhancer also contains binding sites for several other factors in addition to the octamer motif (Figure 5.4). One of these (μB) binds a factor that is found only in B cells, whilst the others (E1–E4) bind ubiquitous factors present in all cell types. As with the octamer, however, deletion or mutation of the binding sites for these ubiquitous factors reduces the activity of the enhancer in B cells, even though the octamer motif is intact (Lenardo *et al.*, 1987). Hence in the heavy-chain gene enhancer B cell-specific activity is achieved by the

Figure 5.4 Structure of the immunoglobulin heavy-chain gene enhancer. Note the binding sites (shaded) for the B cell-specific factors μB and Oct–2 (0) and the E1–E4 sites (unshaded) which bind constitutive factors.

Figure 5.5 Structure of the immunoglobulin light-chain gene enhancer. Note the binding sites (shaded) for the B cell-specific factors NFκB and Oct-2 and the E1–E3 sites (unshaded) which bind constitutive factors.

interaction of B cell-specific factors and ubiquitous factors which bind to adjacent sites in the enhancer.

A similar mixture of binding sites for ubiquitous and B cell-specific proteins is found in the enhancer element of the immunoglobulin κ light-chain gene (Figure 5.5) and is critical for its activity. Thus, for example, the enhancer contains a motif known as E2 which serves as a binding site for two ubiquitously expressed proteins, E12 and E47 (Murre *et al.*, 1989), as well as two other binding sites (E1 and E3) for constitutively expressed factors. The E12 and E47 factors play a critical role in B cell development since inactivation of the E2A gene, which encodes both these factors, results in the arrest of B cell development at an early stage (review: Dorshkind, 1994).

As well as binding sites for such constitutively expressed factors, the light-chain enhancer also contains a binding site for Oct-2 (Nelms and van Ness, 1990) and for another B cell-specific protein, NFκB, which is a heterodimer of two subunits p50 and p65 (reviews: Blank *et al.*, 1992; Grimm and Baeurele, 1993). The binding of NFκB to its binding site in the enhancer plays an especially critical role in the expression of the light-chain gene. Thus mutation or deletion of the NFκB-binding site abolishes the B cell-specific activity of the light-chain enhancer (Lenardo *et al.*, 1987) whilst, like the octamer motif, it can render a heterologous promoter B cell-specific (Pierce *et al.*, 1988).

This suggests, therefore, that binding of NFκB to its site in the enhancer controls the B cell-specific activity of the enhancer. In agreement with this, binding activity can be detected only in B cell lines that express the immunoglobulin κ gene and not in pre-B cells or other cell types which do not. Moreover, treatment of pre-B cells with lipopolysaccharide results in the appearance of NFκB binding activity and concomitantly activates κ gene expression (Atchison and Perry, 1987).

This activation of NFκB by lipopolysaccharide treatment can occur in the presence of the protein synthesis inhibitor cycloheximide. Hence NFκB is present in B cells in an inactive form and can be activated post-translationally without new protein synthesis. Interestingly, this inactive form of NFκB is widely distributed in different cell types and can be activated in both T cells and HeLa cells by treatment with phorbol ester. In these cases NFκB activation does not result in immunoglobulin light-chain gene expression, since the gene has not rearranged and is tightly packed within inactive chromatin, but it does play a role in gene regulation. Thus the activation of NFκB by agents that activate T cells results in the active transcription factor inducing increased expression of cellular genes such as that encoding the interleukin–2α receptor, and is also responsible for the increased activity of the human immunodeficiency virus promoter in activated T cells. NFκB

Figure 5.6 Activation of NFκB by dissociation of the inhibitory protein IκB, allowing NFκB to dimerize, move to the nucleus and switch on gene expression. Note that dissociation of IκB from NFκB is caused by its phosphorylation (P). NFκB is shown as a single factor for simplicity, although it normally exists as a heterodimer of two subunits, p50 and p65.

therefore plays a role not only in B cell-specific gene activity but also in gene activity specific to activated T cells.

Unlike Oct-2, therefore, NFκB is present in all tissues and is activated post-translationally. This occurs, as with HSF, by increased phosphorylation. However, the primary target for phosphorylation is not NFκB but an inhibitory protein associated with it, known as IκB (reviews: Beg and Baldwin, 1993; Gilmore and Morin, 1993). Phosphorylation of IκB results in its dissociation from NFκB, allowing NFκB to move to the nucleus in a manner similar to the nuclear movement of hsp90-free glucocorticoid receptor following steroid treatment (Figure 5.6). For further discussion of the mechanisms activating NFκB, see Section 10.3.3.

In summary, therefore, B cell-specific expression of the immunoglobulin genes is controlled by the B cell-specific activity of their promoters and enhancers, which is in turn controlled primarily by the B cell-specific synthesis of Oct-2 and the B cell-specific activation of NFκB.

5.3 MYO D AND THE CONTROL OF MUSCLE-SPECIFIC GENE EXPRESSION

5.3.1 Identification of MyoD

Probably the most novel approach to the cloning of the gene encoding a transcription factor was taken by Davis *et al.* (1987), who isolated cDNA clones encoding MyoD, a factor that plays a critical role in skeletal muscle-specific gene regulation. They used an embryonic muscle fibroblast cell line known as C3H 10T1/2. Although these cells do not exhibit any differentiated characteristics, they can be induced to differentiate into myoblast cells expressing a number of muscle-lineage genes upon treatment with 5-azacytidine (Constantinides *et al.*, 1977). This agent is a cytidine analogue having a nitrogen instead of a carbon atom at position 5 on the pyrimidine ring, and is incorporated into DNA instead of cytidine. Unlike cytidine, however, it cannot be methylated at this position and hence its incorporation results in demethylation of DNA. As methylation of DNA at C residues is thought to play a critical role in transcriptional silencing of gene expression (reviews: Razin and Cedar, 1991; Latchman, 1995), this artificial demethylation can result in the expression of particular genes that were previously silent.

In the case of 10T1/2 cells, therefore, this demethylation was thought to result in the expression of previously silent, regulatory loci which are necessary for differentiation into muscle myoblasts. Several experiments also suggested that the activation of only one key regulatory locus might be involved. Thus 5-azacytidine induces myoblasts at very high frequency, consistent with the demethylation of only one gene being required, whilst DNA prepared from differentiated cells can also induce differentiation in untreated cells at a frequency consistent with the transfer of only one activated locus.

Hence differentiation is thought to occur via the activation of one regulatory locus (gene X in Figure 5.7) whose expression in turn switches on the expression of genes encoding muscle lineage markers which is observed in the differentiated 10T1/2 cells and thereby induces their differentiation. This suggested that the regulatory locus might encode a transcription factor that switched on muscle-specific gene expression.

To isolate the gene encoding this factor, Davis *et al.* (1987) reasoned that it would continue to be expressed in the myoblast cells but not in the undifferentiated cells. They therefore prepared RNA from the differentiated cells and removed from it by subtractive hybridization all the RNAs that were also expressed in the undifferentiated cells. After various further manipulations to exclude RNAs characteristic of terminal muscle differentiation, such as myosin and others induced non-specifically in all cells by 5-azacytidine, the enriched probe was used to screen a cDNA library prepared from differentiated 10T1/2 cells.

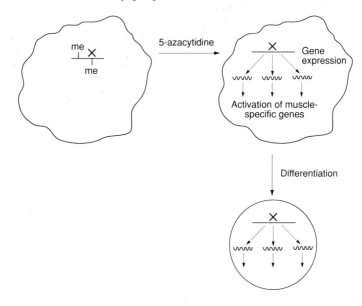

Figure 5.7 Model for differentiation of 10T1/2 cells in response to 5-azacytidine. Activation of a master locus (x) by demethylation allows its product to activate the expression of muscle-specific genes, thereby producing differentiation.

RNA from differentiated 10T1/2 cells

Make radioactive cDNA

Remove RNA which is also found in undifferentiated cells by subtractive hybridization

Remove RNAs for products of terminal muscle differentiation e.g. desmin, myosin heavy and light chains

Remove RNAs induced by 5-azacytidine in cells where myoblasts are not produced

Screen cDNA library to isolate clones corresponding to remaining RNAs

MyoA MyoD MyoH

Figure 5.8 Strategy for isolating the master regulatory locus expressed in 10T1/2 cells after but not before treatment with 5-azacytidine. Subtractive hybridization was used to isolate all RNA molecules that are present in 10T1/2 cells only following treatment with 5-azacytidine. After removal of RNAs for terminal differentiation products of muscle and RNAs induced in non-muscle-producing cells by 5-azacytidine, the remaining RNAs were used to a screen a cDNA library. Three candidates for the master regulatory locus, MyoA, MyoD and MyoH, were isolated in this way.

This procedure (Figure 5.8) resulted in the isolation of three clones, *MyoA*, *MyoD* and *MyoH*, whose expression was specifically activated when 10T1/2 cells were induced to form myoblasts with 5-azacytidine. When each of these genes was artifically expressed in 10T1/2 cells, *MyoA* and *MyoH* had no effect. However, artificial expression of *MyoD* was able to convert undifferentiated 10T1/2 cells into myoblasts (Figure 5.9). Hence expression of *MyoD* alone can induce differentiation of 10T1/2 cells.

The differentiated 10T1/2 cells produced in this manner, like those induced by 5-azacytidine, express a variety of muscle lineage markers and indeed also switch on both *MyoA* and *MyoH* as well as the endogenous MyoD gene itself (Thayer *et al.*, 1989). This suggests that MyoD is a transcription factor which switchs on genes expressed in muscle

Figure 5.9 Test of each of the putative master regulatory loci, *MyoA*, *MyoD* and *MyoH*. Each of the genes was introduced into 10T1/2 cells and tested for the ability to induce the cells to differentiate into muscle cells. Note that, whilst *MyoA* and *MyoH* have no effect, introduction of *MyoD* results in the production of muscle cells that contain the muscle protein myosin. The differentiated muscle cells induced by *MyoD* cease to divide on differentiation, resulting in fewer cells being detectable by staining with crystal violet compared with the *MyoA*- and *MyoH*-treated cells, which continue to proliferate. Hence only *MyoD* has the capacity to cause 10T1/2 cells to differentiate into non-proliferating muscle cells producing myosin, identifying it as a master regulatory locus for muscle differentiation.

cells. In agreement with this, MyoD was shown to bind to a region of the creatine kinase gene upstream enhancer which was known to be necessary for its muscle-specific gene activity (Lassar *et al.*, 1989).

Interestingly, as well as stimulating muscle-specific genes, MyoD also promotes differentiation by modulating gene expression so as to inhibit cellular proliferation, thereby producing the non-dividing phenotype characteristic of muscle cells. Thus MyoD has recently been shown to activate the gene encoding the p21 inhibitor of cyclin-dependent kinases (Halevy *et al.*, 1995). This results in the inhibition of these kinases, whose activity is necessary for cell division (see Section 7.3.2). In addition, MyoD can also repress the promoter of the c-*fos* gene, whose protein product is important for cellular proliferation (see Section 7.2.1), indicating that MyoD can also act by repressing genes whose products are not required in non-dividing muscle cells (Trouche *et al.*, 1993).

Like gene activation by MyoD, repression of the c-*fos* promoter is dependent on DNA binding, which in this case prevents the binding of a positively acting factor to a site known as the serum response element which overlaps the MyoD-binding site in the c-*fos* promoter (Figure 5.10). Obviously, in contrast to its binding to the creatine kinase enhancer, MyoD must bind to its binding site in the c-*fos* promoter in a form that cannot activate transcription. Hence, like the glucocorticoid recep-

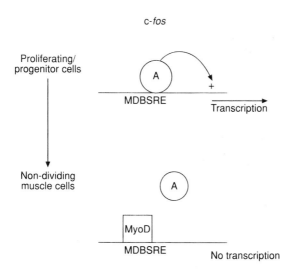

Figure 5.10 MyoD binds to its binding site (MDBS) in the c-*fos* promoter in a configuration that does not activate transcription and prevents binding of an activating factor (A) to the overlapping serum response element (SRE). This therefore results in the repression of c-*fos* transcription in MyoD-expressing muscle cells.

tor (see Section 4.3.4), MyoD can have different effects on gene expression depending on the nature of its binding site.

In both cases, however, DNA binding by MyoD is dependent upon a basic region of the protein which binds directly to the DNA and an adjacent region which can form a helix–loop–helix structure and is essential for dimerization of MyoD. These elements, which are also found in other transcription factors such as the E12 and E47 proteins that bind to the E2 site in immunoglobulin enhancers (Murre *et al.*, 1989; see Section 5.2.2), are discussed further in Section 8.4.

Hence MyoD is a transcription factor whose activation by 5-azacytidine treatment results in the activation of muscle-specific gene expression. Interestingly, the observation that the introduction of MyoD into cells switches on the endogenous *MyoD* gene (see above) suggests that a positive feedback loop normally regulates *MyoD* expression so that, once the gene is initially expressed, expression is maintained, producing commitment to the myogenic lineage (Figure 5.11).

Despite its ability to induce cells in culture to differentiate into muscle cells, however, mice that lack any functional MyoD develop normally and produce skeletal muscle. However, if a second gene encoding the Myf-5 protein is also deleted, no skeletal muscle is formed even though inactivation of Myf-5 alone does not prevent skeletal muscle formation (Rudnicki *et al.*, 1993). Hence either MyoD or Myf-5 alone can induce skeletal muscle formation *in vivo* in the absence of the other, resulting in lack of skeletal muscle only when both proteins are inactive. As expected from this, both Myf-5 and another related protein myogenin are also able to induce the differentiation of 10T1/2 cells into myoblasts

Figure 5.11 Ability of MyoD protein to activate expression of its own gene, creating a positive feedback loop which ensures that, following an initial stimulus, the MyoD protein is continuously produced and hence maintains myoblast differentiation.

and, together with MyoD, these proteins form a family of related transcription factors containing the helix–loop–helix motif which play a key role in controlling muscle differentiation (reviews: Edmondson and Olson, 1993; Olson and Klein, 1994).

5.3.2 Regulation of MyoD

As expected in view of the critical role that MyoD plays in the development of muscle cells, MyoD mRNA is present in skeletal muscle tissue taken from a variety of different sites in the body but is absent in all other tissues including cardiac muscle (Davis *et al.*, 1987; Figure 5.12).

Figure 5.12 Northern blotting experiment to detect the mRNAs encoding MyoA, MyoD and MyoH in different muscle and non-muscle tissues. Note that the MyoD mRNA is present only in skeletal muscle, as expected in view of its ability to produce muscle differentiation, whereas the MyoA and MyoH mRNAs are more widely distributed. nb, newborn; ad, adult; rRNA, the ribosomal RNA control used to show that all samples contain intact RNA.

LIVERPOOL JOHN MOORES UNIVERSITY

Hence, like Oct-2, MyoD mRNA and protein accumulate only in a specific cell type where they are required, and activation of the MyoD gene during myogenesis is likely to be of central importance in switching on the expression of muscle-specific genes.

However, in addition to its control by regulation of its synthesis, MyoD activity also appears to be regulated in another manner. Thus, in 10T1/2 cells, increased expression of MyoD, whether produced by 5-azacytidine treatment or by introduction of the *MyoD* gene, results in the formation of myoblasts. However, this increased expression of MyoD

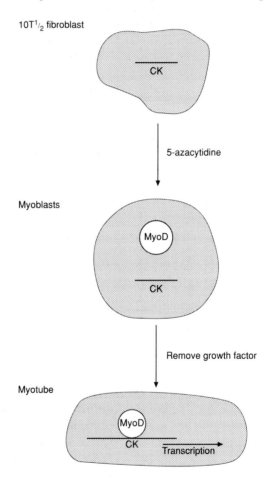

Figure 5.13 Differentiation of 10T1/2 cells into myoblasts by 5-azacytidine and then into myotubes by removal of growth factors. Note that the MyoD-dependent induction of genes encoding terminal differentiation markers such as creatine kinase (CK), which occurs in myotubes, involves no increase in MyoD concentration.

does not result in the formation of fully differentiated muscle cells or in the activation of several genes that can be activated by MyoD *in vitro*, such as creatine kinase. The activation of these products of terminal muscle cell differentiation requires the removal of growth factors from the culture medium of the myoblasts, resulting in cell fusion and the formation of fully differentiated myotubes. Although this process results in switching on of MyoD-dependent genes such as the creatine kinase gene, no further increase in the level of MyoD is observed during this differentiation (Figure 5.13).

Hence a paradox exists, with the MyoD-dependent genes that encode muscle differentiation markers becoming activated only when myotubes are formed even though the MyoD level is the same as that in myoblasts. The explanation of this paradox was provided by the identification of the Id protein (Benezra *et al.*, 1990) which, like MyoD, contains a helix–loop–helix motif but lacks the basic domain mediating DNA binding. Because the helix–loop–helix motif mediates dimerization of proteins containing it, Id can dimerize with other helix–loop–helix proteins such as MyoD and inhibit their DNA binding since the resulting heterodimer lacks the necessary pair of DNA-binding motifs (Figure 5.14).

Interestingly, when 10T1/2-derived myoblasts are induced to form

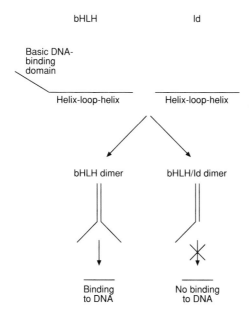

Figure 5.14 Dimerization of functional basic helix–loop–helix proteins (bHLH) with Id. Note that, whilst Id can dimerize with other proteins via the helix–loop–helix domain, it lacks the basic DNA domain and hence the Id-containing heterodimer cannot bind to DNA.

myotubes, Id levels decline, indicating that this second stage of myogenesis is mediated by a decline in the inhibitory protein rather than an increase in the activator, MyoD. The simplest model for the induction of muscle-specific genes during this process would involve increased levels of free MyoD being produced by its release from Id, allowing it to bind to DNA and activate transcription. In fact, however, it appears that the real situation is more complex. Thus high-affinity DNA binding by MyoD to muscle-specific genes such as that encoding creatine kinase requires it to form heterodimers with the ubiquitously expressed helix–

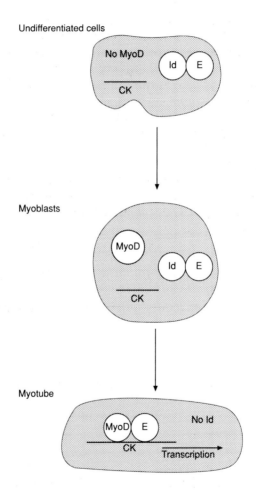

Figure 5.15 Activation of muscle terminal differentiation markers such as creatine kinase (CK) following differentiation of myoblasts to myotubes. The fall in Id concentration when myotubes are formed from myoblasts releases another helix–loop–helix protein (E), which dimerizes with MyoD and activates gene expression.

loop–helix proteins, E12 and E47, which are both encoded by the E2A gene. Thus, in addition to binding to the E2 motif in the immunoglobulin enhancer (Section 5.2.2), these E2A proteins also bind, like MyoD, to the creatine kinase enhancer. However, a dimer of MyoD and an E2A protein binds to this site with much higher affinity than either homo-dimer, and strongly activates gene expression (Lassar *et al.*, 1991). Moreover, the E2A proteins bind to Id with much higher affinity than does MyoD, as predicted for proteins whose activity is regulated by association with Id (Benezra *et al.*, 1990).

These observations suggest a model (Figure 5.15) in which, in myo-blasts, Id is predominantly associated with E2A proteins. On addition of serum, Id levels fall, releasing the E2A proteins, allowing them to heterodimerize with MyoD, leading to gene activation and myotube formation. In agreement with this idea, the over-expression of Id in myoblasts inhibits myotube formation (Jen *et al.*, 1992). Hence, as with the immunoglobulin enhancer, the activation of myotube-specific genes may be controlled by the interplay of tissue-specific and constitu-tive factors, although in the myotube case such factors associate together before DNA binding rather than binding to distinct sites, as in the immunoglobulin enhancer.

The role of inhibitory helix–loop–helix proteins is not confined to myogenesis. Thus the level of Id also declines during differentiation of early embryonic and erythroid cells, whilst the product of the *emc* gene, which regulates neurogenesis in *Drosophila*, also contains a helix–loop–helix motif and lacks a basic DNA-binding domain (review: Jones, 1990).

In summary, therefore, regulation of MyoD appears to combine the types of regulation we have noted in Oct-2 and NFκB, being controlled both by its increased synthesis in myoblasts and by its increased activity mediated by changes in protein–protein interaction during the forma-tion of myotubes.

5.4 REGULATION OF YEAST MATING TYPE

5.4.1 Yeast mating type

In the preceding sections, we have discussed some of the control mechanisms that regulate the expression of different genes in the vast array of differentiated cells found in higher eukaryotes. Clearly such expression of different genes within different cells of the same organism does not occur in single-celled eukaryotes such as yeast, where the organism consists of only a single cell. None the less, similar regulatory processes operate in these organisms. Thus, in yeast, two distinct mating types exist, known as a and α, which must fuse together to create a diploid cell. These two mating types each express distinct genes encoding products such as pheromones or pheromone receptors which are involved in mating. Individual yeasts of different mating type can thus be regarded as analogous to the different types of differentiated cell in a higher eukaryote, although they are in fact distinct individual organisms of different phenotype.

In some yeast strains, known as heterothallic strains, the two mating types are entirely separate, as in higher organisms. We shall discuss, however, the situation in homothallic yeast strains where each cell is capable of switching its mating type from a to α or vice versa, and behaves following switching exactly as does any other cell with its new mating type (reviews: Herskowitz, 1989; Dolan and Fields, 1991).

This switching process is a very precise one (Figure 5.16), occurring only in the G1 phase of the cell cycle prior to cell division and only in a mother cell that has already produced a daughter by budding. The daughter itself does not switch until it has grown and produced a daughter itself. This process is thought to be a consequence of the need rapidly to produce diploid progeny, switching from a to α or vice versa, allowing this to occur amongst the progeny of a single cell without the need for contact with another strain of different mating type. This process is of interest in terms of gene regulation from two points of view. Firstly, how does a cell switch from a to α or vice versa, and secondly,

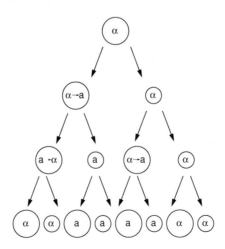

Figure 5.16 Mating-type switching in yeast. In every generation the larger mother cell, which has produced a smaller daughter, switches its genotype from a to α or vice versa.

how are genes expressed only in a or α cells regulated? These two aspects will be discussed in turn.

5.4.2 Control of mating-type switching

Genetic analysis of homothallic yeast strains allowed the demonstration that whether a cell was a or α in phenotype was controlled by a single gene locus on chromosome 3, known as MAT (mating type). If this transcriptionally active locus contained an a gene, the cell was of a mating type; if it contained an α gene, then the cell was of α mating type. In addition, however, yeast cells also contain transcriptionally silent copies of both the a and α genes elsewhere on chromosome 3 at the HML and HMR loci. The change in mating type occurs via a cassette mechanism in which one of the silent copies replaces the active gene at the MAT locus, changing the mating type (Figure 5.17). This process is controlled by an endonuclease which is the product of the HO (homo-thallism) gene and which makes a double-stranded cut at the MAT locus, initiating switching.

At first sight, this process involving DNA cutting by an endonuclease appears to have little relevance to the study of gene regulation by transcription factors. A more detailed consideration of the process reveals, however, a central role for transcription factors. Thus the activity of the HO endonuclease is controlled by transcriptional regula-tion of its corresponding gene. A number of regulatory processes operate on this gene, ensuring that it is transcribed only in the G1 phase of the cell cycle and only in mother cells and not in daughter cells, resulting in the observed restriction of the switching process to the G1 phase in mother cells.

Genetic analysis has allowed the identification of a number of differ-ent loci that regulate HO expression. Thus five SWI genes necessary for the expression of HO have been defined, whilst six SIN loci which inhibit

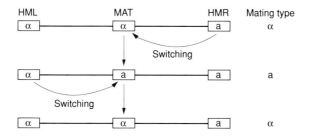

Figure 5.17 Mechanism of mating-type switching involving the movement of an a or α gene from the inactive HML or HMR loci to the active MAT locus.

HO expression have also been identified. It is likely that these genes encode transcription factors that activate or repress HO gene expression in particular situations. Thus the SWI4 and SWI6 gene products have been shown to be necessary for the formation of a protein complex on DNA sequences upstream of the HO gene (review: Andrews and Mason, 1993). This complex binds to a DNA sequence in HO which is necessary for its expression in G1 and which can confer this expression pattern on an unrelated gene. It is likely, therefore, that the SWI4 and SWI6 gene products play a critical role in the activation of HO expression and hence of switching only in the G1 phase of the cell cycle. Similarly, the activation of the HO gene only in mother cells and not in daughter cells is controlled by the opposing actions of the SIN3 gene product, which inhibits HO expression, and of SWI5, which antagonizes the action of SIN3 and activates HO transcription (Sternberg *et al.*, 1987; Figure 5.18). The SWI5 protein contains zinc finger motifs characteristic of DNA-binding transcription factors (see Section 8.3.1) and binds specifically to sequences in the HO promoter involved in mother/daughter control (Stillman *et al.*, 1988).

Therefore, even though mating-type switching is actually brought about by the action of an endonuclease, the activity of the endonuclease and hence switching is regulated at the level of gene transcription by several interacting control mechanisms.

5.4.3 Gene regulation by the a and α gene products

Once mating-type switching has determined whether the mating-type locus contains an α or a gene, the expression of this gene in turn regulates the activation or repression of genes specific to a or α cells. This is achieved at the level of transcription, the a or α genes encoding transcription factors that can activate or repress gene expression. Inter-

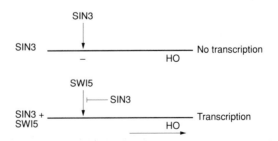

Figure 5.18 Anatagonistic action of the SWI5 and SIN3 gene products on HO transcription. Note that when SWI5 is present it antagonizes the inhibitory effect of SIN3 and HO transcription occurs.

estingly, DNA sequence analysis of these genes (Shepherd *et al.*, 1984) has revealed that they contain a motif with strong homology to the homeobox that is shared by a number of different transcription factors involved in regulating embryonic development in *Drosophila* and other higher eukaryotes (Figure 5.19; see Section 6.2) and which mediates DNA binding by the factors that contain it (see Section 8.2).

Hence the a and α products control mating type by modulating the transcription of genes whose products are required for the a or α phenotype. This is not achieved, however, by the a product switching on a-specific genes and the α product switching on α-specific genes. Rather, the products of the a and α loci interact to regulate the a- and α-specific genes as well as the expression of genes whose products are required in both mating types but not in diploid cells (Figure 5.20) (review: Dolan and Fields, 1991).

Genetic analysis of this process has indicated that the a-specific genes are constitutively active but are repressed by one of the products of the α

| Ftz | Ser | Leu | Ser | Glu | Arg | Gln | Ile | Lys | Ile | Trp | Phe | Gln | Asn | Arg | Arg | Met | Lys | Ser | Lys |
| α 2 | Ser | Leu | Ser | Arg | Ile | Gln | Ile | Lys | Asn | Trp | Val | Ser | Asn | Arg | Arg | Arg | Lys | Glu | Lys |

Figure 5.19 Relationship of the yeast mating-type protein α2 and the homeobox of the *Drosophila* Fushi-tarazu protein. Boxes indicate identical amino acids.

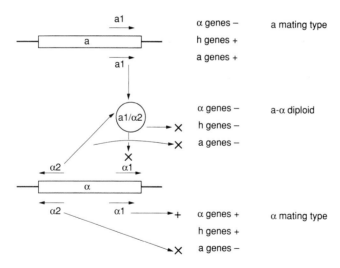

Figure 5.20 Regulatory actions of the a and α gene products. Note that, whilst the α1 product is required for the activation of the α-specific genes, the α2 product inhibits the constitutively active a-specific genes. The interaction of a1 and α2 proteins, which occurs in a–α diploids where both are present, inhibits the expression of the haploid-specific genes (h) as well as that of α1, thereby indirectly inhibiting the α-specific genes.

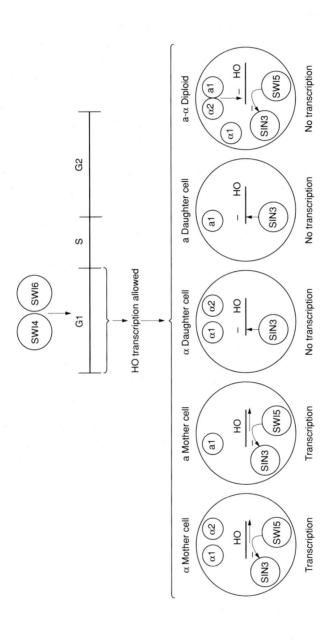

Figure 5.21 Transcription of the HO gene. Note that transcription is allowed only in the G1 phase of the cell cycle when the SWI4 and SWI6 gene products are present and in mother cells where SWI5 inhibits the action of SIN3. Transcription of HO is also inhibited in a–α diploids by the action of the a1 and α2 proteins.

gene, α 2. Hence they are active in a cells where α 2 is absent but not in α cells where it is present. Similarly, they are repressed in the presence of α2 in a–α diploid cells where the mating genes must be repressed since the cells have already mated and fused to give a diploid.

In contrast to the constitutive activity of the a-specific genes, the α-specific genes are inactive in the absence of the other product of the α gene, α1 and hence are inactive in a cells where it is absent but active in α cells where it is present. These genes are also inactive, however, in diploids where α1 is present. This paradox is explained by the fact that when both α2 and the product of the a gene, a1, are present, they co-operate to repress both the α1 gene (hence indirectly preventing α gene expression) and also the genes whose expression is required in both mating types but not in non-mating diploids. Interestingly, one of the genes repressed by the a1–α2 complex is that encoding HO which, although required for switching in both a and α cells is not required in diploids. This gene is therefore repressed by the a1–α2 complex and its G1-specific expression is hence subject to multiple controls, preventing expression in both haploid daughter cells and in diploids (Figure 5.21). A summary of the regulatory processes acting on the HO gene is given in Figure 5.22.

Hence genetic studies have indicated that gene expression is regulated by the action of the α1 activator protein and the a1 and α2 repressor proteins. These studies have now been supplemented by a more detailed molecular analysis of the a and α proteins and their interaction with the promoters of the genes they regulate (review: Dolan and Fields, 1991). Thus it has been shown that the constitutive activity of the a-specific genes is controlled by the binding of a constitutively expressed transcription factor, MCMI (also known as GRM (general regulator of mating type) or PRTF (pheromone receptor transcription factor)) to a sequence upstream of the a-specific genes known as the P box. Interestingly, MCMI is also required for transcription of

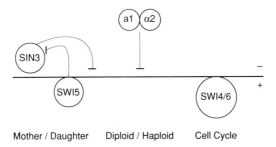

Figure 5.22 Summary of the mechanisms regulating HO transcription involving the negative actions of SIN3 and the a1–α2 complex and the positive actions of SWI5 and the SWI4/6 complex.

the α-specific genes and is capable of binding to a sequence upstream of these genes known as the P' box, which is related to but distinct from the P box. After binding to this P' box, however, transcriptional activation does not occur unless the α1 protein is present. This protein forms a complex with MCMI, and transcriptional activation occurs (Figure 5.23).

Therefore the distinction between the a- and α-specific genes is controlled by sequences in their promoters that allow transcriptional activation of the a-specific genes following binding of MCMI alone, with transcriptional activation of the α-specific genes occurring only after binding of the MCMI–α1 complex and not following binding of MCMI alone. A possible mechanism for this effect has been suggested by the finding that binding of MCMI to the P box produces a conformational change in the structure of the protein which does not occur after binding to the P' box. A similar conformational change in MCMI is observed, however, following binding to the P' box in the presence of the α1 protein. This conformational change is postulated to expose a transcriptional activation domain on MCMI, allowing it to activate transcription. Hence the difference in gene activation following binding to the P and P' boxes is controlled by the fact that MCMI alone can undergo the necessary conformational change after binding to the P box but requires α1 to do so following binding to the P' box.

This model therefore explains the dependence of the α-specific genes

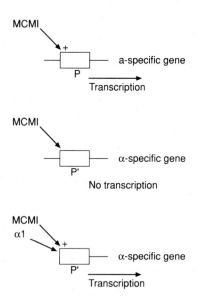

Figure 5.23 Stimulation of the a-specific genes by binding of MCMI to the P box and of the α-specific genes by binding of MCMI and α1 to the P' box. Note that MCMI cannot activate the α-specific genes via P' in the absence of α1.

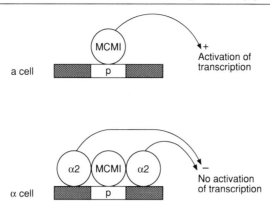

Figure 5.24 Inhibition of the a-specific genes in α cells by binding of α2 to sites flanking the binding site for MCMI.

on the α1 factor for their expression. A further aspect of the model is required to explain how the constitutively active a-specific genes are repressed in α mating-type cells. This occurs because the P box in these genes is flanked by binding sites for the α2 protein. One molecule of the α2 protein binds on either side of the MCMI protein bound to the P box and prevents the transcription of the a-specific genes (Figure 5.24). This inhibition of the activity of MCMI does not take place by α2 preventing the binding of MCMI to the P box. Thus the binding of MCMI is actually enhanced by the presence of α2, with the two proteins binding co-operatively to the DNA. Hence α2 does not appear to act by interfering with the action of positively acting factors. Rather, it appears to be able to inhibit transcription directly by interacting with the basal transcriptional complex (Herschbach *et al.*, 1994), as occurs for the thyroid hormone receptor in the absence of thyroid hormone (see Section 4.3.4). This latter possibility is supported by the finding that the presence of the α2–MCMI complex prevents transcription from the promoter directed by any other unrelated activation sequences placed upstream of the α2–MCMI binding site, as would be expected if it directly inhibits the activity of the basal transcriptional complex, rendering it non-responsive to upstream activating proteins (see Section 9.3 for discussion of the mechanisms by which transcription factors repress gene expression).

The differential activity of the a- and α-specific genes is therefore controlled by the constitutive factor MCMI and its interaction with the cell type-specific factors α1 and α2 (Figure 5.25).

In addition to its role as a repressor of the a-specific genes, α2 also plays a key role in the repression in diploid cells of genes such as HO, whose activity is required for mating in general and which are therefore expressed in both a and α cells but not in non-mating a–α diploids. This

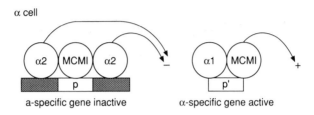

Figure 5.25 Regulation of a- and α-specific genes by the constitutive factor MCMI and the mating-type factors α1 and α2.

repression is dependent upon the association of the α2 protein with the a1 protein. This association requires the DNA-binding domain of the a1 protein and both the DNA-binding domain and a C-terminal region of the α2 protein (Stark and Johnson, 1994) and results in high-affinity co-operative binding of the two proteins upstream of the haploid-specific gene promoters. The inhibitory activity of the α2 protein (see above) then results in the repression of transcription (Figure 5.26).

Interestingly, a distinct region at the N-terminus of the α2 protein mediates its interaction with the MCMI protein (see above), producing similar co-operative binding of this complex to the promoters of the a-specific genes. Thus the α2 protein, which in isolation binds DNA non-specifically and with low affinity, is directed to distinct DNA-binding sites by interaction with either the MCM1 or a1 proteins. Interaction with MCMI results in dimerization of α2, which then binds to sites in the a-specific genes. In contrast, interaction with a1 results in the formation of an a1–α2 heterodimer with a different DNA-binding specificity, leading to binding to sites in the haploid-specific genes (Figure 5.26). Hence, the α2 protein can be regarded as a generalized transcriptional repressor which is targeted to different genes in different situations, depending on the other proteins present (Smith and Johnson, 1992; Goutte and Johnson, 1993).

It is clear, therefore, that the products of the a and α mating-type genes act to regulate gene expression via complex interactions both with

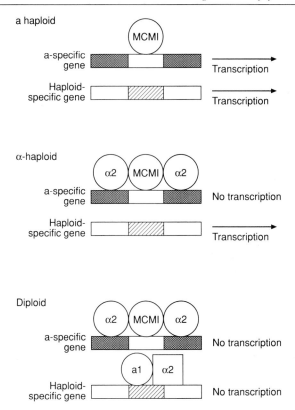

Figure 5.26 Inhibition by the α2 gene product. In a cells, α2 is absent, so both the a-specific genes and the haploid-specific genes are transcribed. In α haploid cells, α2 can repress the a-specific genes but cannot bind to haploid-specific genes, which are therefore transcribed. In diploid cells, however, α2 forms a heterodimer with the a1 factor, which can bind to the haploid-specific genes and repress their transcription.

themselves and with constitutive factors. A detailed analysis of this process and the control of switching itself has been made possible by the combination of the genetic techniques available in yeast, and the molecular techniques described in Chapter 2. The possibility that these processes, which are understood in such detail, may serve as a model for similar regulatory processes in higher organisms is discussed in the next section.

5.4.4 Relevance of the yeast mating-type system to higher organisms

At first sight, a process such as mating-type switching, which involves DNA rearrangement, is of apparently little relevance to higher organisms where, with very few exceptions such as the immunoglobulin genes (see Section 5.2.1), DNA rearrangements have not been shown to be involved in developmental processes. The more detailed discussion of this process given above illustrates, however, a number of possible events in this process that are applicable to higher organisms.

Thus both the action of the a and α gene products and the regulation of HO transcription involve the complex interaction of different antagonistic or synergistic transcription factors. Many of these interactions, such as the difference in binding specificity of homo- and heterodimers or the inhibition of gene activation by one factor that is mediated via a second factor, both of which have been extensively studied in this system, have now been shown to have counterparts in higher organisms (see Section 10.3.3). Similarly, the yeast transcription factors themselves show considerable similarity in structure to factors present in higher organisms. Thus MCMI shows considerable homology to the mammalian serum response transcription factor (SRF) and can bind to sites within mammalian genes that bind this factor (Passmore *et al.*, 1989) whilst, as discussed above (Section 5.4.3), the a and α gene products contain the homeobox also found in many developmentally regulated transcription factors in higher eukaryotes. Moreover, the interaction between MCMI and the α2 protein in yeast is paralleled

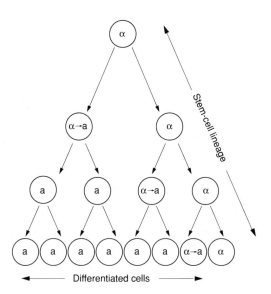

Figure 5.27 Model for the generation of a stem-cell lineage, producing differentiated cells by a system based on α to a mating-type switching. In this model, α represents the stem cell and a the differentiated cell, and it is assumed that, unlike the yeast mating-type system, the α to a switch is irreversible.

by that between the SRF factor and proteins of the homeobox family in mammalian and *Drosophilia* cells (Grueneberg *et al.*, 1992).

Hence the ability to study transcription factors and their interactions both genetically and biochemically in yeast has led to the accumulation of considerable information of potential relevance to gene control in higher eukaryotes.

Interestingly, Herskowitz (1985) has drawn an analogy between mating-type switching and lineage in higher eukaryotes. Thus, in the switching system illustrated in Figure 5.16, if α represents a stem cell and a is a differentiated cell derived from it, then, if it is assumed that, unlike the yeast situation, switching from α to a is irreversible, the model lineage illustrated in Figure 5.27 is obtained. In this model, the stem cell is continually dividing, with one daughter maintaining the stem cell lineage whilst the other differentiates. This type of system is commonly found in higher organisms, being used in the development of a number of different cell types and organs.

Even though such a lineage is unlikely to have a DNA rearrangement as its basis, the simple model of change in expression from one state to another provided by the yeast system can be used to generate models of how differentiation might occur in higher organisms.

5.5 CONCLUSIONS

In this chapter we have discussed the regulation of cell type-specific transcription via the activity of specific transcription factors that are synthesized or are present in an active form only in the particular cell type where the gene(s) they regulate are active. By interacting both with each other and with constitutively expressed factors, these cell type-specific factors control the specific transcription pattern of the genes that are dependent upon them.

In some cases, such as Oct-2 or NFκB, these factors simply regulate one or more genes that are active in the cell types where they are present. In other cases, such as MyoD or the yeast mating-type proteins, however, the factor is capable of directly or indirectly switching on all the genes whose expression is characteristic of the differentiated cell, and its expression is therefore sufficient to produce the differentiated cell phenotype.

In lower eukaryotes such as yeast, therefore, the switching on of such a transcription factor in a particular cell will result in the appropriate differentiated phenotype. Although this is the case in higher eukaryotes also, such activation must occur at a particular stage of development and in a particular group of adjacent cells so that a tissue or organ forms at the appropriate time and in the appropriate place. The role of transcription factors in this regulation of gene expression during development is discussed in the next chapter.

REFERENCES

Andrews, B. J. and Mason, S. W. (1993). Gene expression and the cell cycle; a family affair. *Science* **261**, 1543–1544.

Atchison, M. L. and Perry, R. P. (1987). The role of the κ enhancer and its binding factor NF-κB in the developmental regulation of κ gene transcription. *Cell* **48**, 121–128.

Beg, A. A. and Baldwin, A. S. (1993). The IκB proteins: multifunctional regulators of Rel/NFκB transcription factors. *Genes and Development* **7**, 2064–2070.

Benezra, R., Davis, R. L., Lockshon, D., Turner, D. L. and Weintraub, H. (1990). The protein Id: a negative regulator of helix–loop–helix DNA binding proteins. *Cell* **61**, 49–59.

Blank, V., Kowitsky, P. and Israel, A. (1992). NFκB and related proteins: Rel/dorsal homologies meet ankyrin-like repeats. *Trends in Biochemical Sciences* **17**, 135–140.

Clerc, R. G., Corcoran, L. M., LeBowitz, J. H., Baltimore, D. and Sharp, P. A. (1988). The B cell specific Oct-2 protein contains POU box and homeo box type domains. *Genes and Development* **2**, 1570–1581.

Constantinides, P. G., Jones, P. A. and Gevers, W. (1977). Functional striated muscle cells from non-myoblast precursors following 5-azacytidine treatment. *Nature* **267**, 364–366.

Davis, H.L., Weintraub, H. and Lassar, A.B. (1987). Expression of a single transfected cDNA converts fibroblasts to myoblasts. *Cell* **51**, 987–1000.

Dolan, J.K. and Fields, S. (1991). Cell type-specific transcription in yeast. *Biochimica et Biophysica Acta* **1088**, 155–169.

Dorshkind, K. (1994). Transcriptional control points during lymphopoiesis. *Cell* **79**, 751–753.

Edmondson, D. F. and Olson, E. N. (1993). Helix–loop–helix proteins as regulators of muscle-specific transcription. *Journal of Biological Chemistry* **268**, 755–758.

Garcia, J. V., Bich-Thuy, L., Stafford, J. and Queen, C. (1986). Synergism between immunoglobulin enhancers and promoters. *Nature* **322**, 383–385.

Gellert, M. (1992). Molecular analysis of V(D)J recombination. *Annual Review of Genetics* **22**, 425–446.

Gerster, T., Mathias, P., Thali, M., Jariciny, J. and Schaffner, W. (1987). Cell type-specificity of the immunoglobulin heavy chain gene enhancer. *EMBO Journal* **6**, 1323–1330.

Gilmore, T. D. and Morin, P. J. (1993). The IκB proteins: members of a multi-functional family. *Trends in Genetics* **9**, 427–433.

Goutte, C. and Johnson, A. D. (1993). Yeast a1 and α2 homeodomain proteins form a DNA binding activity with properties distinct from that of either protein. *Journal of Molecular Biology* **233**, 359–371.

Grimm, S. and Baeurele, P. A. (1993). The inducible transcription factor NFκB: structure–function relationship of its protein subunits. *Biochemical Journal* **290**, 297–308.

Grueneberg, D. A., Natesan, S., Alexandre, C. and Gilman, M. Z. (1992). Human and *Drosophila* homeodomain proteins that enhance DNA binding activity of serum response factor. *Science* **257**, 1089–1095.

Gstaiger, M., Knoepfel, L., Georgiev.O., Schaffner, W. and Hovens, C. M. (1995). A B-cell co-activator of octamer-binding transcription factors. *Nature* **373**, 360–362.

Halevy, O., Novitch, B. G., Spicer, D. B., Skapek, S. X., Rhee, J., Hannon, G. J., Beach, D. and Lassar, A. B. (1995). Correlation of terminal cell cycle arrest of skeletal muscle with induction of p21 by MyoD. *Science* **267**, 1018–1021.

Herschbach, B. M., Arnaud, M. B. and Johnson, A. D. (1994). Transcriptional repression directed by the yeast α2 protein *in vitro*. *Nature* **370**, 309–311.

Herskowitz, L. (1985). Master regulatory loci in yeast and lambda. *Cold Spring Harbor Symposia* **50**, 565–574.

Herskowitz, I. (1989). A regulatory hierachy for cell specialization in yeast. *Nature* **342**, 749–757.

Jen, Y., Weintraub, H. and Benezra, R. (1992). Overexpression of Id protein inhibits the muscle differentiation program: *in vivo* association of Id with E2A proteins. *Genes and Development* **6**, 1466–1479.

Jones, N. (1990). Transcriptional regulation by dimerization: two sides to an incestuous relationship. *Cell* **61**, 9–11.

Lassar, A. B., Buskin, J. N., Lockshon, D., Davis, R. L., Apone, S., Hauschka, S. D. and Weintraub, H. (1989). MyoD is a sequence-specific DNA binding protein requiring a region of *myc* homology to bind to the muscle creatine kinase enhancer. *Cell* **58**, 823–831.

Lassar, A. B., Davis, R. L., Wright, W. E., Kadesch, T., Murre, C., Voronova, A., Baltimore, D. and Wentraub, H. (1991). Functional activity of myogenic HLH proteins requires hetero-oligomerization with E12/E47-like proteins *in vivo*. *Cell* **66**, 305–315.

Latchman, D. S. (1995). *Gene Regulation: A Eukaryotic Perspective*, 2nd edn. London: Chapman and Hall.

Lenardo, M., Pierce, J. W. and Baltimore, D. (1987). Protein binding sites in Ig gene enhancers determine transcriptional activity and inducibility. *Science* **236**, 1573–1577.

Mignotte, V., Wall, L., de Boer, E., Grosveld, F. and Romeo, P. H. (1989). Two tissue-specific factors bind the erythroid promoter of the human porphobillinogen deaminase gene. *Nucleic Acids Research* **17**, 37–54.

Murre, C., McCaw, P. S. and Baltimore, D. (1989). A new DNA binding and dimerization motif in immunoglobulin enhancer binding, *daughterless*, MyoD and myc proteins. *Cell* **56**, 777–783.

Nelms, K. and van Ness, B. (1990). Identification of an octamer binding site in the human κ light chain enhancer. *Molecular and Cellular Biology* **10**, 3843–3846.

Olson, E. N. and Klein, W. H. (1994). bHLH factors in muscle development: dead lines and commitments, what to leave in and what to leave out. *Genes and Development* **8**, 1–8.

Parslow, T. G., Blair, D. C., Murphy, W. J. and Granner, D. K. (1984). Structure of the 5' ends of immunoglobulin genes: a novel conserved sequence. *Proceedings of the National Academy of Sciences USA* **81**, 2650–2654.

Passmore, S., Elble, R. and Tye, B-K. (1989). A protein involved in minichromosome maintenance in yeast binds a transcriptional enhancer conserved in eukaryotes. *Genes and Development* **3**, 921–935.

Pevny, L., Simon, M. C., Robertson, E., Klein, W. H., Tsai, S. F., D'Agati, V., Orkin, S. H. and Constantini, F. (1991). Erythroid differentiation in chimaeric mice blocked by a targetted mutation in the gene for transcription factor GATA–1. *Nature* **349**, 257–260.

Pfisterer, P., Annweiler, A., Ullmer, C., Corcoran, L.M. and Wirth, T. (1994). Differential transactivation potential of Oct-1 and Oct-2 is determined by additional B cell specific activities. *EMBO Journal* **13**, 1654–1663.

Pierce, J. W., Lenardo, M. and Baltimore, D. (1988). Oligonucleotide that binds nuclear factor NF-κB acts as a lymphoid specific and inducible enhancer element. *Proceedings of the National Academy of Sciences USA* **85**, 1482–1486.

Razin, A. and Cedar, H. (1991). DNA methylation and gene expression. Microbiological Reviews 55, 451–458.

Reitman, M. and Felsenfeld, G. (1988). Mutational analysis of the chicken beta globin enhancer reveals two positive acting domains. *Proceedings of the National Academy of Sciences USA* **85**, 6267–6271.

Roitt, I. M., Brostoff, J. and Male, D. K. (1993). *Immunology*. St Louis, MO: Mosby.

Rudnicki, M. A., Schnegelsberg, P. N. J., Stead, R. H., Brown, T., Arnold,

H. H. and Jaenisch, R. (1993). MyoD or Myf-5 is required for the formation of skeletal muscle. *Cell* **75**, 1351–1359.

Shepherd, J. C. W., McGinnis, W., Carrasco, A. E., De Robertis, E. M. and Gehring, W. J. (1984). Fly and frog homoeo domains show homologies with yeast mating type regulatory loci. *Nature* **310**, 70–71.

Singh, H., Sen, R., Baltimore, D. and Sharp, P. A. (1986). A nuclear factor that binds to a conserved sequence motif in transcriptional control elements of immunoglobulin genes. *Nature* **319**, 154–158.

Smith, O. L. and Johnson, A. D. (1992). A molecular mechanism for combinational control in yeast: MCMI protein sets the spacing and orientation of the homeodomains of an α2 dimer. *Cell* **68**, 133–142.

Stark, M. R. and Johnson, A. D. (1994). Interaction between two homeodomain proteins is specified by a short C-terminal trail. *Nature* **371**, 429–432.

Sternberg, P. W., Stern, M. J., Clark, I. and Herskowitz, I. (1987). Activation of the yeast HO gene by release from multiple negative controls. *Cell* **48**, 567–577.

Stillman, D. J., Bonkier, A. T., Seddan, A., Groenheat, G. and Nasmylh, K. A. (1988). Characterization of a mother cell transcription factor involved in mother cell specific transcription of the yeast HO gene. EMBO *Journal* **7**, 485–494.

Strubin, M., Newell, J. W. and Matthias, P. (1995). OBF–1, a novel B cell-specific coactivator that stimulates immunoglobulin promoter activity through association with octamer-binding proteins. *Cell* **80**, 497–506.

Thayer, M. J., Tapscott, S. J., Davis, R. L., Wright, W. E., Lassar, A. B. and Weintraub, H. (1989). Positive autoregulation of the myogenic determination gene MyoD1. *Cell* **58**, 241–248.

Trouche, D., Grigoriev, M., Lenormard, J. C., Robin, P., Leibovitch, S. A., Sassone-Corsi, P. and Havel-Bellar, A. (1993). Repression of c-*fos* promoter by MyoD on muscle cell differentiation. *Nature* **363**, 79–82.

Wirth, T., Staudt, L. and Baltimore, D. (1987). An octamer oligonucleotide upstream of a TATA motif is sufficient for lymphoid specific promoter activity. *Nature* **329**, 174–178.

CHAPTER SIX

Transcription factors and developmentally regulated gene expression

6.1 DEVELOPMENTALLY REGULATED GENE EXPRESSION

As discussed in Chapter 5, the ability to carry out genetic experiments in yeast resulted in the isolation of genes whose mutagenesis affects cell type-specific gene regulation, allowing the role of their protein products in this process to be elucidated. Clearly such an approach could be equally valuable in understanding the regulation of gene expression during development in which the processes of cell type-specific gene expression discussed in Chapter 5 are integrated and co-ordinated so that each cell type and tissue arises in the right place at the right time. It is evident, however, that this cannot be done in the single-celled yeast. Rather, these studies have focused on the fruit fly *Drosophila melanogaster*, which is well characterized genetically and has a short generation time, facilitating its study.

A large number of mutations that affect the development of this organism have been isolated and their corresponding genes named on the basis of the observed phenotype of the mutant fly (reviews: Ingham, 1988; Lawrence and Morata, 1994). Thus mutations in the so-called homeotic genes result in the transformation of one particular segment of the body into another; mutations in the *Antennapedia* gene, for example, cause the transformation of the segment that normally produces the antenna into one that produces a middle leg (Figure 6.1). Similarly, mutations in genes of the gap class result in the total absence of particular segments; mutations in the *Knirps* gene, for

Figure 6.1 Effect of the homeotic mutation *Antennapedia*, which produces a middle leg (B) in the region that would contain the antenna of a normal fly (A). aI, aII, aIII, 1st, 2nd and 3rd antennal segments; ar, arista; ta, tarsus; ti, tibia; fe, femur; ap, apical bristle.

example, result in the absence of most of the abdominal segments, although the head and thorax develop normally.

The products of genes of this type therefore play critical roles in *Drosophila* development. The products of the gap genes, for example, are necessary for the production of particular segments, whilst the homeotic gene products specify the identify of these segments. Given that these processes are likely to require the activation of genes whose protein products are required in the particular segment, it is not surprising that many of these genes have been shown to encode transcription factors. Thus the *Knirps* gene product and that of another gap gene, *Kruppel*, contain multiple zinc finger motifs characteristic of DNA-binding transcription factors and can bind to DNA in a sequence-specific manner (Rosenberg *et al.*, 1986). Similarly, the *tailless* gene, whose product plays a key role in defining the anterior and posterior regions of the *Drosophila* embryo has been shown to be a member of the steroid receptor gene superfamily discussed in Section 4.3 (Pignoni *et al.*, 1990).

It is clear, therefore, that the genes identified by mutation as playing a role in *Drosophila* development can encode several different types of transcription factors. However, of the first 25 such genes that were cloned, allowing a study of their protein products, well over half (15) contain a motif known as the homeobox (Gehring *et al.*, 1994) which was originally identified in the homeotic genes of *Drosophila*. The evidence that these homeobox proteins modulate transcription in *Drosophila* and the ways in which they do so are discussed in Section 6.2.

Following the successful use of genetic analysis to isolate the *Drosophila* homeobox genes, attempts have been made to isolate homologues of these genes in other organisms, less amenable to genetic analysis, by using hybridization with the DNA of the *Drosophila* genes. Accordingly, Section 6.3 discusses the isolation of homeobox genes in other organisms, including mammals, by this and other means, as well as the evidence that their protein products are involved in developmental gene regulation in these organisms.

6.2 THE HOMEOBOX-CONTAINING GENES OF DROSOPHILA

6.2.1 The homeobox

When the first homeotic genes were cloned, it was found that they shared a region of homology approximately 180 base pairs long and therefore capable of encoding 60 amino acids, which was flanked on either side by regions that differed dramatically between the different genes. This region was named the homeobox (review: Gehring *et al.*, 1994). Subsequently, the homeobox was shown to be present in many other *Drosophila* regulatory genes. These include the *Fushi-tarazu* gene (Ftz), which is a member of the pair-rule class of regulatory loci whose mutation causes alternate segments to be absent, and the *engrailed* gene (Eng), which is a member of the class of genes whose products regulate segment polarity. The close similarity of the homeoboxes encoded by the homeotic genes *Antennapedia* and *Ultrabithorax* and that encoded by the Ftz gene is shown in Figure 6.2.

The presence of this motif in a large number of different regulatory genes of different classes strongly suggested that it was of importance in their activity. The evidence that the homeobox-containing proteins are transcription factors whose DNA-binding activity is mediated by the homeobox is discussed in the next section (reviews: Hayashi and Scott, 1990; Gehring *et al.*, 1994).

Block 1 (residues 1–30):

Antp	Arg	Lys	Arg	Gly	Arg	Gln	Thr	Tyr	Thr	Arg	Tyr	Gln	Thr	Leu	Glu	Leu	Glu	Lys	Glu	Phe	His	Phe	Asn	Arg	Tyr	Leu	Thr	Arg	Arg	Arg
Ubx		Arg																				Thr		His						
Ftz	Ser			Thr																						Ile				

Block 2 (residues 31–60) — Helix–Turn–Recognition helix region:

		Helix								Turn			Recognition helix																	
Antp	Arg	Ile	Glu	Ile	Ala	His	Ala	Leu	Cys	Leu	Thr	Glu	Arg	Gln	Ile	Lys	Ile	Trp	Phe	Gln	Asn	Arg	Arg	Met	Lys	Trp	Lys	Lys	Glu	Asn
Ubx				Met																						Leu				Ile
Ftz			Asp			Asn			Ser		Ser															Ser			Asp	Arg

Figure 6.2 Amino acid sequences of several *Drosophila* homeodomains, showing the conserved helical motifs. Differences between the sequences of the Ubx and Ftz homeodomains from that of Antp are indicated; a blank denotes identity in the sequence. The helix–turn–helix region is indicated.

6.2.2 The homeobox proteins as transcription factors

The first indication that the homeobox proteins were indeed transcription factors came from the finding that the homeobox was also present in the yeast mating type a and α gene products. Thus, as discussed in Section 5.4, these proteins are known to be transcription factors which regulate the activity of a- and α-specific genes, hence suggesting, by analogy, that the *Drosophila* proteins also fulfilled such a role.

Direct evidence that this is the case is now available from a number of different approaches. Thus it has been shown that many of these proteins bind to DNA in a sequence-specific manner, as expected for transcription factors (Hoey and Levine, 1988). Moreover, binding of a specific homeobox protein to the promoter of a particular gene correlates with the genetic evidence that the protein regulates expression of that particular gene. For example, the Ultrabithorax (Ubx) protein has been shown to bind to specific DNA sequences within its own promoter and in the promoter of the *Antennapedia* gene, in agreement with the genetic evidence that Ubx represses *Antennapedia* expression (Figure 6.3).

The ability of the homeobox-containing proteins to bind to DNA is directly mediated by the homeobox itself. Thus, if the homeobox of the

Figure 6.3 Assay of protein binding to a DNA fragment from the *Antennapedia* gene promoter (Antp) or a control fragment of plasmid DNA (pUC) using protein extracts from *Escherichia coli* which have been genetically engineered to express the *Drosophila* Ubx protein (a) or protein extracts from control *E. coli* not expressing Ubx (b). Note the specific binding of Ubx protein to the *Antennapedia* DNA fragment.

Antennapedia protein is synthesized in isolation either in bacteria or by chemical synthesis, it is capable of binding to DNA in the identical sequence-specific manner characteristic of the intact protein. The structural features of the homeodomain which allow it to do this are discussed in Section 8.2.

Although DNA binding is a prerequisite for the modulation of transcription, it is necessary to demonstrate that the homeobox proteins do actually affect transcription following such binding. In the case of the Ubx protein, this was achieved by Krasnow *et al.* (1989), who showed that co-transfection of a plasmid expressing Ubx with a plasmid in which the *Antennapedia* promoter drives a marker gene resulted in the repression of gene expression driven by the *Antennapedia* promoter. Hence the observed binding of Ubx to the Antp promoter (see above) results in down-regulation of its activity, in agreement with the results of genetic experiments.

Most interestingly, the Ubx expression plasmid was able to up-regulate activity of its own promoter in co-transfection experiments, this ability being dependent on the previously defined binding sites for Ubx within its own promoter. Similarly, although Ubx normally has no effect on expression of the alcohol dehydrogenase (Adh) gene, it can stimulate the Adh promoter following linkage of the promoter to a DNA sequence containing multiple binding sites for Ubx (Krasnow *et al.*, 1989). Hence

Figure 6.4 Effect of Ubx on various marker genes with or without binding sites (hatched boxes) for the Ubx protein. Note that Ubx can stimulate its own promoter which contains a Ubx-binding site and this effect is abolished by deleting the Ubx-binding site. Similarly, the alcohol dehydrogenase (Adh) gene, which is normally unaffected by Ubx, is rendered responsive to Ubx stimulation by addition of Ubx-binding sites. In contrast, the *Antennapedia* promoter, which also contains Ubx-binding sites, is repressed by Ubx. Hence binding of Ubx can activate or repress different promoters.

a homeobox protein can produce distinct effects following binding, Ubx activating its own promoter and a hybrid promoter containing Ubx-binding sites but repressing the activity of the Antp promoter (Figure 6.4).

A similar transcriptional activation effect of DNA binding has been demonstrated for the Fushi-tarazu (Ftz) protein. This protein binds specifically to the sequence TCAATTAAATGA. As with Ubx, linkage of this sequence to a marker gene confers responsivity to activation by Ftz, such activation being dependent upon binding of Ftz to its target sequence, a one-base-pair change which abolishes binding, also abolishing the induction of transcription (Jaynes and O'Farrell, 1988; Figure 6.5).

These co-transfection studies, carried out by introducing target DNA into cultured cells along with another DNA molecule expressing the homeobox protein, have been supplemented by studies using *in vitro* transcription assays. Thus, if the Ubx promoter linked to a marker gene is added to a suitable cell-free extract, transcription of the marker gene driven by the promoter can be observed. Addition of the purified homeodomain protein even-skipped (eve) to this extract inhibits Ubx promoter activity, however, and this inhibition is dependent upon binding sites for the eve protein within the Ubx promoter (Biggin and Tjian, 1989). Such findings parallel the ability of a vector expressing eve to repress the Ubx promoter following co-transfection into cultured cells and the genetic evidence which originally led to the definition of eve as a repressor of Ubx (Figure 6.6). Like the thyroid hormone receptor (Section 4.3.4) and the yeast α2 protein (Section 5.4.3), the eve protein acts as a direct transcriptional repressor, inhibiting the activity of the basal transcriptional complex (Johnson and Krasnow, 1992) (for further discussion of directly acting transcriptional repressors see Section 9.3.2).

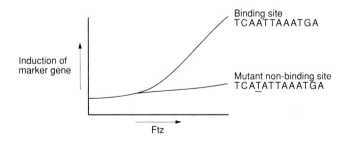

Figure 6.5 Effect of expression of the Ftz protein on the expression of a gene containing its binding site, or a mutated binding site containing a single base-pair change which abolishes binding of Ftz.

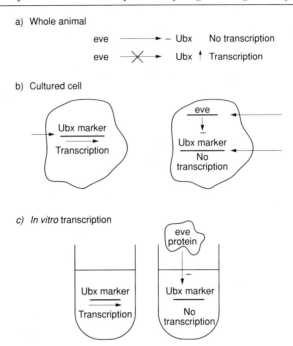

Figure 6.6 Inhibitory effect of the eve protein on expression of the Ubx gene. This inhibitory effect can be observed in the whole animal, where mutation of the eve gene enhances Ubx expression (a), in cultured cells, where introduction of a plasmid expressing the eve gene represses a co-transfected Ubx promoter driving a marker gene (b) and in a test tube *in vitro* transcription system, where addition of purified eve protein represses transcription of a marker gene driven by the Ubx promoter (c).

Hence, in the case of eve, a direct link exists between the genetic data, the results of co-transfections into cells and the *in vitro* data. The fact that the results of experiments in whole organisms or intact cells can be directly reproduced in cell extracts indicates that the homeodomain proteins function directly as transcription factors to control development. Moreover, by showing that these factors can act in a relatively simple *in vitro* system, they provide a means of analysing this effect.

A variety of evidence is therefore available which allows the conclusion that the homeobox proteins are transcription factors that exert their effects on development by activating or repressing the activity of target genes. What remains to be elucidated is the manner in which such relatively simple effects on gene expression can regulate the enormously complicated process of development. The progress that has been made in this area is discussed in the next section.

6.2.3 Homeobox transcription factors and the regulation of development

At first sight it is remarkably difficult to understand how the relatively simple process of transcriptional regulation by homeobox proteins could in turn control development. Although this process is of course not yet fully understood, a number of mechanisms have been defined which indicate ways in which the complex regulatory networks needed to regulate development might be built up.

One such finding has already been discussed in the previous section. Thus, in a number of cases, it has already been shown that a single factor can repress some target genes whilst activating others, thereby increasing the range of effects mediated by one factor. The Ubx protein, for example, can activate transcription from its own promoter whilst repressing that of the Antp gene (Krasnow *et al.*, 1989). At least in neuronal cells, this effect appears to operate by the Ubx protein binding to the Antp promoter and preventing the binding to the same site of the Antp factor itself, which would normally activate the promoter. Hence this represents an example of gene repression by interfering with the binding of another activating factor, rather than by direct repression, as in the case of the eve protein discussed above (see Section 9.3 for discussion of the mechanisms of negative regulation).

Whatever its mechanism, the activation and repression of different promoters by the Ubx protein has important consequences in terms of the control of development. Thus the ability of Ubx to induce its own transcription provides a mechanism for the long-term maintenance of Ubx gene expression during development, since once expression has been switched on and some Ubx protein made, it will induce further transcription of the gene via a simple positive feedback loop even if the factors that originally stimulated its expression are no longer present (Figure 6.7). This long-term maintenance of Ubx expression is essential since, if the Ubx gene is mutated within the larval imaginal disc cells which eventually produce the adult fly, the cells that would normally produce the haltere (balancer) will produce a wing instead. Although these cells are known to already be committed to form the adult haltere at the larval stage, the continued expression of the Ubx gene is essential to maintain this commitment and allow eventual overt differentiation (see Hadorn (1968) for a review of imaginal discs and their role in *Drosophila* development).

Similarly, the inhibition of the Antp gene by Ubx indicates that the homeobox factors do not act simply by activating the transcription of genes for structural proteins or enzymes required in particular cell types. Although this must be one of their functions, they can also regulate the transcription of each other, creating the potential for

Figure 6.7 The stimulatory effect of the Ubx protein on the transcription of its own gene ensures that, once Ubx gene transcription is initially switched on by an activator protein (A), transcription will continue even if the activator protein is removed.

regulatory networks. We have already seen how the eve protein can repress Ubx gene transcription whilst Ubx in turn represses the Antp promoter. Since Antp has been shown to stimulate both its own promoter and that of Ubx (Winslow *et al.*, 1989), this creates the possibility of complex interactions in which the synthesis of one particular factor at a particular time will create changes in the levels of numerous other factors and ultimately result in the activation or repression of numerous target genes.

As well as regulating the expression of other transcription factor genes in this manner, the homeobox proteins evidently also regulate the expression of genes whose protein products are required in a particular cell type, such as the cell-surface adhesion molecules (review: Edelman and Jones, 1993). Hence by regulating both each other's expression and that of non-homeobox target genes both positively and negatively, the homeobox transcription factors can create regulatory networks of the type necessary for the control of development.

The effects that we have described so far involve the activity of single homeobox factors affecting the expression of other genes. In addition, however, it is also possible for the same target gene to be regulated by multiple homeobox factors, with the effects observed with one factor being different, depending on whether or not another factor is also present.

In one such mechanism the DNA-binding specificity of one factor is altered in the presence of another factor. Thus several homeobox proteins, such as Ubx and Antp, bind to the same DNA sequences when tested in isolation *in vitro* (Hoey and Levine, 1988) yet, parodoxically, the effects of mutations that inactivate the genes encoding each of these proteins are different, indicating that they cannot substitute for one another. Similarly, *in vivo*, Ubx can bind to a site in the promoter of the decapentaplegic (dpp) gene and activate its expression, whereas Antp cannot do so.

This paradox is explained by the presence in the dpp promoter of a binding site for another homeobox protein, extradenticle (exd), which lies adjacent to the site to which Ubx binds. The exd protein interacts with the Ubx protein and modifies its DNA-binding specificity so it can bind to the dpp gene promoter and activate its expression (Figure 6.8) (Chan *et al.*, 1994; Van Dijk and Murre, 1994). This interaction with exd is dependent on the homeobox and C-terminal region of Ubx, which have several amino acid differences from the corresponding region of Antp. Because of these differences, Antp does not interact with exd and hence cannot bind to the site in the dpp gene promoter.

This interaction clearly parallels the process in which the yeast α2 homeodomain protein binds to different binding sites following interaction with either MCM1 or the a1 protein (see Section 5.4.3). This suggests that the modification of the DNA-binding specificity of a homeobox factor by interaction with various other factors represents an evolutionarily conserved strategy for targeting these factors to different genes in different situations.

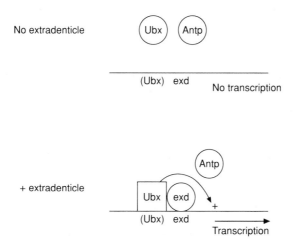

Figure 6.8 The exd protein interacts with the Ubx protein to allow it to bind with high affinity to its potential binding site in the dpp promoter (indicated as (UbX)) and activate its expression. In contrast, the Antp protein cannot interact with exd and so does not bind to the dpp promoter.

As well as interactions that alter DNA binding, other interactions operate at a functional level, with combinations of factors either synergizing with each other or interfering with one another so that functional effects are observed with combinations of factors which are not observable with either factor alone.

For example, it has been shown that the homeobox proteins engrailed (Eng), Fushi-tarazu (Ftz), paired (prd) and zerknult (zen) can all bind to the sequence TCAATTAAAT (Hoey and Levine, 1988). When plasmids expressing each of these genes are co-transfected with a target promoter carrying multiple copies of this binding site, the Ftz, prd and zen proteins can activate transcription of the target promoter (Jaynes and O'Farrell, 1988; Han *et al.*, 1989). In contrast, the Eng protein has no effect on the transcription of such a promoter. It does, however, interfere with the ability of the activating proteins to induce transcription, presumably by blocking the binding of the activating factor. Thus, for example, whilst Ftz can stimulate the target promoter when co-transfected with it, it cannot do so in the presence of Eng (Jaynes and O'Farrell, 1988). Hence the expression of Ftz alone in a cell would activate particular genes, whereas its expression in a cell also expressing engrailed would not have any effect (Figure 6.9).

Similar types of interaction can also take place between different positively acting factors binding to the same site. Thus, although, as noted above, the Ftz, prd and zen products can all activate transcription of a target promoter when transfected alone, this effect is relatively small, producing only approximately two-fold activation. In contrast, much larger effects can be obtained by activating the target promoter with two of these proteins in combination, producing 10–20-fold activation, or by all three activators together, producing 400-fold activation of the target (Han *et al.*, 1989; Figure 6.10).

This synergistic stimulation of the target promoter by factors that

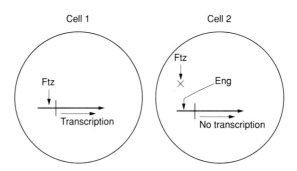

Figure 6.9 Blockage of gene induction by Ftz in cells expressing the engrailed (Eng) protein, which binds to the same sequence as Ftz but does not activate transcription.

Transfected DNA				Effect on target promoter
Ftz	prd	zen	Eng	
–	–	–	–	–
+	–	–	–	+
+	+	–	–	+++
+	–	+	–	+++
+	+	+	–	+++++
+	–	–	+	–

Figure 6.10 Activation of a target promoter containing multiple copies of the identical binding site for the Ftz, prd and zen proteins when it is co-transfected with plasmids expressing each of these proteins. Note the synergistic activation of the promoter when more than one of the activator proteins is co-transfected and the inhibitory effect of the Eng protein, which binds to the same site as the other proteins but inhibits gene expression.

Figure 6.11 Model for the synergistic activation of gene expression by the various homeobox proteins which have the same DNA-binding site as illustrated in Figure 6.10. Gene activation is suggested to be enhanced when different factors (A, B, C) are bound to the multiple copies of their binding site (BS) present in the target promoter compared with the level of activation when the same factor (A) is bound to each of the binding sites.

bind the same sequence can be explained on the basis of the observation that the target promoter contains multiple copies of the binding site for these factors. It has been suggested, therefore (Han *et al.*, 1989), that gene activation proceeds via a multi-switch mechanism in which much stronger activation is produced by binding of different factors at each binding site than by the binding of multiple copies of the same factor (Figure 6.11).

Hence the complex effects of single homeobox factors on the expression of other genes can be rendered still more complex by means of synergistic or inhibitory effects of combinations of factors creating effects that would not be obtained with a single factor alone. Indeed,

Figure 6.12 Model system producing variation in the spatial expression of three different responder genes along the length of an embryo, as indicated by the open or shaded circles. All the responder genes are activated by the activator A and inhibited by the two repressors, R1 and R2, each of which is expressed in different but overlapping regions of the embryo, indicated by the solid, open or hatched boxes. The three genes differ, however, in their relative affinity for the activators and repressors, as indicated. This results in variation between the genes in whether they are activated or repressed in the presence of a particular combination of activator and repressor molecules in each part of the embryo, as indicated by the open circles (no expression) or the shaded circles (expression). Thus, for example, in the second circle both A and R1 are present, so genes 1 and 3 are not expressed (since R1 binds more strongly than A to these genes) while gene 2 is expressed (since A binds more strongly than R1 to this gene).

such interactions of different factors can be used to generate models that predict complex spatial distributions of responder gene activity in response to relatively simple expression patterns of homeobox protein distribution. One such model (Jaynes and O'Farrell, 1988; Figure 6.12) is based on the interaction of activator and repressor molecules which bind to the same binding site in the manner of the Ftz and Eng products. By assuming that target genes vary in the affinity of their binding sites for an activator and two repressor molecules whose areas of expression are overlapping but not identical, it is possible to generate different patterns of responder gene activity in each cell type, depending on which particular factors are present (Figure 6.12).

Hence the activation and repression of target genes by different homeobox factors both alone and in combination can generate complex overlapping patterns of target gene expression of the sort that must occur in development. In the model described above and in our discussion so far, however, it has been assumed that a homeobox factor is either present in a particular cell or is entirely absent. In fact, however, a further level of complexity exists since many homeobox factors are not expressed in a simple on–off manner but rather show a concentration gradient ranging from high levels in one part of the embryo via intermediate levels to low levels in another part. For example, the Bicoid protein (Bcd), whose absence leads to the development of a fly without head and thoracic structures, is found at high levels in the anterior part of the embryo and declines progressively posteriorly, being absent in the posterior one-third of the embryo (Figure 6.13; Driever and Nusslein-Volhard, 1988a).

Most interestingly, genes that are activated in response to Bicoid contain binding sites in their promoters that have either high affinity or low affinity for the Bicoid protein. If these sites are linked to a marker gene, it can be demonstrated that genes with low-affinity binding sites are activated only at high concentrations of Bicoid and are therefore expressed only at the extreme anterior end of the embryo. In contrast, genes that have higher-affinity binding sites are active at much lower protein concentrations and will be active both at the anterior end and more posteriorly. Moreover, the greater the number of higher-affinity

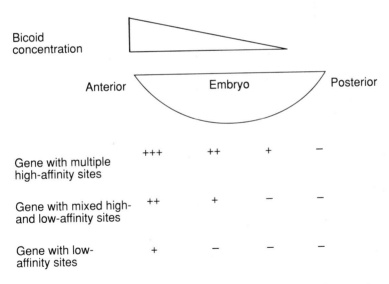

Figure 6.13 The gradient in Bicoid concentration from the anterior to the posterior point of the embryo results in Bicoid-dependent genes with only low-affinity binding sites for the protein being active only at the extreme anterior part of the embryo, whereas genes with high-affinity binding sites are active more posteriorly. Note that, in addition to the different posterior boundaries in the expression of genes with high- and low-affinity binding sites, genes with high-affinity binding sites will be expressed at a higher level than those with low-affinity binding sites at any point in the embryo.

sites the greater the level of gene expression that will occur at any particular point in the gradient (Driever *et al.*, 1989; Figure 6.13).

The gradient in Bicoid expression can be translated, therefore, into the differential expression of various Bicoid-dependent genes along the anterior part of the embryo. Each cell in the anterior region will be able to 'sense' its position within the embryo and respond by activating specific genes. One of the genes activated by Bicoid is the homeobox-containing segmentation gene *Hunchback* (Tautz, 1988). In turn this protein regulates the expression of the gap genes *Kruppel* and *giant* (Struhl *et al.*, 1992). All four of these proteins then act on the eve gene,

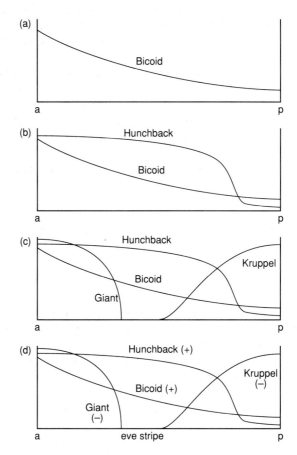

Figure 6.14 Model illustrating how the concentration gradients of the activators Bicoid and Hunchback and the repressors Kruppel and Giant produce a stripe of eve gene expression. Note that the Bicoid gradient (a) affects Hunchback expression (b), which in turn affects Giant and Kruppel expression (c). Eve gene activation (+) by Hunchback and Bicoid, and its repression (−) by Giant and Kruppel, then produces a specific stripe or region of the embryo in which eve is expressed (d). a, anterior; p, posterior.

with Bicoid and Hunchback activating its expression whilst Kruppel and giant repress it. The concentration gradients of these four factors thus result in the spatial localization of eve gene expression in a defined region of the embryo where it exerts its inhibitory effects on gene expression (Small *et al.*, 1991; Figure 6.14). Hence the gradient in Bicoid gene expression results in changes in the expression of other genes encoding regulatory proteins, leading to the activation of regulatory networks of the type discussed above.

The Bicoid factor therefore has all the properties of a morphogen whose concentration gradient determines position in the anterior part of the embryo. This idea is strongly supported by the results of genetic experiments in which the Bicoid gradient was artificially manipulated, cells containing artificially increased levels of Bicoid assuming a phenotype characteristic of more anterior cells and vice versa (Driever and Nusslein-Volhard, 1988b).

The studies discussed in this section have indicated, therefore, how by various means, such as affecting each other's expression and that of other target genes both positively and negatively, interacting with each other both synergistically and antagonistically, and by affecting gene expression in a concentration-dependent manner, the homeobox transcription factors can be used to build up the complex regulatory networks that are necessary for the regulation of *Drosophila* development.

6.3 HOMEOBOX-LIKE GENES IN OTHER ORGANISMS

6.3.1 Homeobox-containing genes

The critical role played by the homeobox genes in the regulation of *Drosophila* development suggests that they may play a similar role in other organisms. Indeed, we have already seen that the a and α mating-type gene products which determine the mating type in yeast contain homeoboxes (Section 5.4). Similarly, homeobox-containing genes play a key role in floral development in plants (review: Weigel and Meyerowitz, 1994), whilst in the nematode *Caenorhabditis elegans* homeoboxes have been identified in several genes whose mutation affects development, such as the *mec-3* gene which controls the terminal differentiation of specific sensory cells (Way and Chalfie, 1988).

As in *Drosophila*, studies in the nematode have been facilitated by the availability of well characterized mutations affecting development, allowing the corresponding genes to be isolated and the homeobox identified. In higher organisms, where such genetic evidence was unavailable, numerous investigators have used Southern blot hybridization with labelled probes derived from *Drosophila* homeoboxes in an attempt to identify homeobox-containing genes in these species. Thus, for example, Holland and Hogan (1986) used a probe from the *Antennapedia* homeobox to identify homeobox genes in a wide range of species, including not only other invertebrates, such as the molluscs, but also chordates, such as the sea urchin, and vertebrates including the mouse (Figure 6.15). Subsequent studies have resulted in the identification of a large number of different homeobox-containing genes from a wide variety of organisms, including both mouse and human, and many of these genes have been isolated and their DNA sequence obtained (reviews: Kenyon, 1994; Krumlauf, 1994).

It is clear from these studies that homeobox-containing genes are not confined to invertebrates such as *Drosophila* but are found also in vertebrates, including mammals such as mouse and humans. Interestingly, this evolutionary conservation is not confined to the homeobox portion of these genes. Thus homologues of individual homeobox genes

Figure 6.15 Southern blot of mouse DNA hybridized with a probe from the *Drosophila Antennapedia* gene (A), a mouse *Antennapedia*-like gene (M) and mouse ribosomal DNA (R). Note the presence of DNA fragments which hybridize to both *Antennapedia*-like DNAs but not to ribosomal DNA, and which represent *Antennapedia*-like sequences in the mouse genome.

of *Drosophila*, such as e*ngrailed* and *deformed*, have been identified in mouse and humans, the fly and mammalian proteins showing extensive sequence homology which extends beyond the homeobox to include other regions of the proteins.

Moreover, the similarity between the *Drosophila* and mammalian systems extends to the manner in which the homeobox-containing genes are organized in the genome. Thus in both *Drosophila* and mammals these genes are organized into clusters containing several homeobox-containing genes, with homologous genes in the different organisms occupying equivalent positions in the clusters. For example, in a detailed comparison of the genes in the *Drosophila Bithorax* and *Antennapedia* complexes with those of one mouse homeobox gene com-

Figure 6.16 Comparison of the *Bithorax–Antennapedia* complex on *Drosophila* chromosome 3 with the Hox b complex on mouse chromosome 11. Individual genes are indicated by open boxes. Note that each gene in the *Drosophila* complex is most homologous to the equivalent gene in the mouse complex, as indicated by the vertical lines. The *Drosophila* Abd-A, Ubx and Antp genes are too closely related to each other to be individually related to a particular mouse gene but are most closely related to the Hox b6, b7 and b8 genes, which occupy the equivalent positions in the Hox b cluster as indicated by the brackets. The two alternative nomenclatures for mouse Hox genes are indicated. Note that, in moving from the left to the right of the mouse complex, the genes are expressed progressively earlier in development, have a more anterior boundary of expression and a greater responsiveness to retinoic acid (RA).

plex Hox b (Hox 2), Graham *et al.*, (1989) showed that the first gene in the mouse complex, Hox b-9 (2.5) was most homologous to the first gene in the *Drosophila Bithorax* complex, Abd-B, and so on across the complex (Figure 6.16). Hence both the homeobox genes and their arrangement are highly conserved in evolution, the common ancester of mammals and insects having presumably possessed a similar cluster of homeobox-containing genes.

Most interestingly, in both *Drosophila* and mammals, the position of a gene within a cluster is related to its expression pattern during embryogenesis. Thus, in the mouse Hoxb cluster discussed above, all the genes are expressed in the developing central nervous system of the embryo. However, in moving from the 5′ to the 3′ end of the cluster (i.e. from Hox b-9 (2.5) to Hox b-1 (2.9) in Figure 6.16), each successive gene is expressed earlier in development and also displays a more anterior boundary of expression within the central nervous system (Figures 6.17 and 6.18). Similar expression patterns have also been observed in *Drosophila*, where each successive gene in the *Bithorax* and *Antennapedia* clusters is expressed more anteriorly and affects progressively more anterior segments when it is mutated (Harding *et al.*, 1985).

In the case of the mouse genes a possible molecular mechanism for this differential expression pattern is provided by the finding that genes in the 3′ half of the Hox b cluster are activated in cultured cells by treatment with low levels of retinoic acid, whereas genes in the 5′ half of the cluster require much higher levels of retinoic acid for their activation (review: Langston and Gudas, 1994). Considerable evidence exists that retinoic acid can act as a morphogen in vertebrate development and it has been suggested that a gradient of retinoic acid concentration may exist across the developing embryo (review: Brickell and Tickle, 1989). Hence the observed difference in expression of the Hox b genes could be controlled by a retinoic acid gradient in a manner similar to the differential activation of Bicoid-dependent genes in the Bicoid concentration gradient observed in *Drosophila* (Figure 6.19). In turn, the Hox b genes, like their *Drosophila* counterparts, would switch on other genes required in cells at particular positions in the embryo, accounting for the morphogenetic effects of retinoic acid.

This process of gene activation by retinoic acid is, of course, related to the induction of gene expression in response to treatment with steroid hormones discussed in Section 4.3. Thus retinoic acid functions by binding to specific receptors which are members of the steroid–thyroid hormone receptor superfamily and which in turn bind to specific sequences within retinoic acid-responsive genes, activating their expression (review: Evans, 1988). Hence the activation of regulatory genes and the initiation of a regulatory cascade can be achieved by a modification of the processes mediating inducible gene expression discussed in Chap-

2.5 2.4 2.2

Figure 6.17 Comparison of the expression pattern of the Hox b9 (2.5), b8 (2.4) and b6 (2.2) genes in the 12.5-day mouse embryo. The top panel shows *in situ* hybridization with the appropriate gene probe to a section of the entire embryo, whilst the middle row shows a high-power view of the region in which the anterior limit of gene expression occurs. In these panels, which show the sections in bright field, hybridization of the probe and therefore gene expression is indicated by the dark areas. In the lower panel, which shows the same area in dark field, hybridization is indicated by the bright areas. Note the progressively more anterior boundary of expression of Hox b6 (2.2) compared with Hox b8 (2.4) and Hox b9 (2.5), and compare with their positions in the Hox b (Hox 2) complex in Figure 6.16.

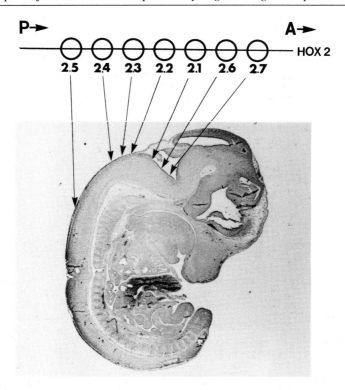

Figure 6.18 Summary of the anterior boundary of expression of the genes in the Hox b (2) complex indicated on a section of a 12.5-day mouse embryo and compared with the position of the gene in the Hox b (2) cluster. Note the progressively more anterior boundary of expression from the 5' to the 3' end of the Hox b (2) cluster.

ter 4. In agreement with this idea, the treatment of mouse embryos with retinoic acid results, for example, in changes in the expression pattern of the Hox b1 gene, which contains a retinoic acid response element in its 3' regulatory region. Moreover, inactivation of this element, so that it no longer binds the retinoic acid receptors, abolishes expression of Hox b1 in the neuroectoderm of the early embryo, providing direct evidence that the retinoic acid response element is necessary to produce the expression pattern of this gene observed in the developing embryo (Marshall *et al.*, 1994).

It is clear, therefore, that homologues of the *Drosophila* homeobox-containing genes do exist in mammals and other vertebrates and that they are expressed in specific cell types in the early embryo. This suggests that they play a similar role to their *Drosophila* counterparts in regulating development by affecting the transcription of other genes. In confirmation of this, a number of cases have been reported in which the inactivation of an individual mouse Hox gene produces developmental

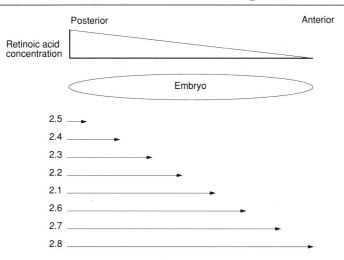

Figure 6.19 Model for the progressively more anterior expression of the genes in the Hox b (2) cluster in which expression is controlled by a posterior to anterior gradient in retinoic acid concentration and the increasing sensitivity to induction by retinoic acid which occurs from the 5′ to the 3′ end of the cluster. Thus, because genes at the 3′ end of the cluster are inducible by very low levels of retinoic acid, they will be expressed in anterior points of the embryo where the retinoic acid level will be too low to induce the genes at the 5′ end of the cluster, which require a much higher level of retinoic acid to be activated.

abnormalities, many of which show the transformations of specific body regions that are characteristic of such mutations in *Drosophila* (see Section 6.1) (review: Krumlauf, 1994).

6.3.2 Homeobox-containing POU proteins

As discussed above, the homeobox-containing genes were first identified in *Drosophila* and only subsequently in other organisms. The reverse is true, however, for another set of transcription factors which possess a homeobox as part of a much larger motif and which were first identified in mammalian cells. Thus, as discussed in Section 5.2.2, the octamer-binding transcription factors Oct-1 and Oct-2 play an important role in regulating the expression of specific genes, such as those encoding histone H2B, the SnRNA molecules and the immunoglobulins. Similarly, the transcription factor Pit-1, which binds to a sequence two bases different from the octamer sequence, plays a critical role in pituitary-specific gene expression (Section 1.2.3).

When the genes encoding these factors were cloned, they were found to share a 150–160-amino-acid sequence, which was also found in the protein encoded by the nematode gene *unc*-86, whose mutation affects sensory neuron development (Finney *et al.*, 1988). This common POU (Pit-Oct-Unc) domain contains both a homeobox sequence and a second conserved domain, the POU-specific domain (Figure 6.20) (reviews: Verrijzer and Van der Vliet, 1993; Wegner *et al.*, 1993).

Interestingly, whilst the homeoboxes of the different POU proteins are closely related to one another (53 of 60 homeobox residues are the same in Oct-1 and Oct-2, and 34 of 60 in Oct-1 and Pit-1), they show less similarity to the homeoboxes of other mammalian genes lacking the POU-specific domain, sharing at best only 21 of 60 homeobox residues. Hence they represent a distinct class of homeobox proteins containing both a POU-specific domain and a diverged homeodomain.

As with the *Drosophila* homeobox proteins, however, the isolated homeodomains of the Pit-1 and Oct-1 proteins are capable of mediating sequence-specific DNA binding in the absence of the POU-specific domain. The affinity and specificity of binding by such an isolated homeodomain is much lower, however, than that exhibited by the intact POU domain, indicating that the POU-specific domain plays a critical role in producing high-affinity binding to specific DNA sequences. Hence the POU homeodomain and the POU-specific domain form two parts of a DNA-binding element, which are held together by a flexible linker sequence. The structure of the POU domain and the manner in which it binds to DNA are discussed further in Section 8.2.

In addition to its role in DNA binding, the POU domain also plays a critical role in several other features of the POU proteins which are not found in the simple homeobox-containing proteins. For example, the ability of both Oct-1 and Oct-2 (like NF1—see Section 3.3.2) to stimulate DNA replication as well as transcription is also a property of the isolated POU domains of these factors (Verrijzer *et al.*, 1990). Similarly, the ability of Oct-1 and not Oct-2 to interact with the herpes simplex virus trans-activator protein Vmw65 is controlled by a single difference in the homeodomain region of the POU domains in the two proteins. Thus the replacement of a single amino acid residue in the homeodomain of Oct-2 with the equivalent amino acid of Oct-1 allows Oct-2 to interact with Vmw65, which is normally a property only of Oct-1 (Lai *et al.*, 1992) (Figure 6.21). In contrast, however, it should be noted that differences in the ability of Oct-1 and Oct-2 to activate different octamer-containing enhancers (see Section 5.2.2) are controlled by regions outside the POU domain of the molecule which contain the activation domains.

As in the case of the homeobox-containing proteins, the POU proteins appear to play a critical role in the regulation of developmental gene expression and in the development of specific cell types. Thus the *unc*-86 mutation in the nematode results, for example, in the lack of touch receptor neurons or male-specific cephalic companion neurons (Desai

Figure 6.20 Amino acid sequences of the POU proteins. The homeodomain and the POU-specific domain are indicated. Solid boxes indicate regions of identity between the different POU proteins. The final line shows a consensus sequence obtained from the four proteins.

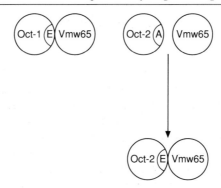

Figure 6.21 Alteration of an alanine residue (A) in the homeodomain of Oct-2 to the glutamic acid residue (E), found at the equivalent position in the homeodomain of Oct-1, allows Oct-2 to interact with the herpes simplex virus trans-activator Vmw65, which is normally a property of Oct-1 only.

et al., 1988), indicating that this POU protein is required for the development of these specific neuronal cell types. Similarly, inactivation of the gene encoding Pit-1 leads to a failure of pituitary gland development, resulting in dwarfism in both mice and humans (review: Voss and Rosenfeld, 1992).

Following the initial identification of the original four POU factors, a number of other members of this family have been described, both in mammals and other organisms such as *Drosophila*. Like the original factors, these novel POU proteins also play a critical role in the regulation of developmental gene expression. Thus, for example, the Drosophila POU protein drifter (CFla) has been shown to be of vital importance in the development of the nervous system (Anderson *et al.*, 1995), whilst mutations in the gene encoding the Brn–4 factor appear to be the cause of the most common form of deafness in humans (de Kok *et al.*, 1995). Moreover, all the novel POU domain-containing genes isolated by He *et al.* (1989) from the rat, on the basis of their containing a POU domain, are expressed in the embryonic and adult brain, suggesting a similar role for these proteins in the regulation of neuronal-specific gene expression. Such a close connection of POU proteins and the central nervous system is also supported by studies using the original POU domain genes, which revealed expression in the embryonic brain even in the case of Oct-2, which had previously been thought to be expressed only in B cells (He *et al.*, 1989).

It is clear, therefore, that, like the homeobox proteins, POU proteins occur in a wide variety of organisms and play an important role in the regulation of gene expression in development. Moreover, these proteins may be of particular importance in the development of the central nervous system.

6.4 CONCLUSIONS

The process of development is clearly immensely complex. The modulation of gene expression during this process involves many of the mechanisms that have been discussed in previous chapters; some genes must become and remain active in all cell types (Chapter 3), others must be induced in response to specific signals (Chapter 4) whilst others must become and remain active in a particular type of differentiated cell (Chapter 5). In turn such processes must be co-ordinated both temporally and spatially so that each organ forms at the right time and in the right place.

The insights provided by the genetic identification of the *Drosophila* transcription factors that control this process have allowed some understanding of how this might be achieved. Thus gradients of transcription factors such as Bicoid, which are laid down in the egg, can be translated into differential gene activity and hence produce specific cell types at particular points in the embryo. Similarly, retinoic acid can activate the expression of specific genes such as the mammalian homeobox-containing genes in a similar manner to the induction of gene expression in response to steroid hormones discussed in Chapter 4. Hence a gradient in a chemical such as retinoic acid can be used to determine position and initiate a regulatory cascade via its ability to activate regulatory genes differentially.

It is clear, therefore, that transcription factors play a critical role in the regulation of development and that gradients in these factors or other substances may allow positional information to be determined. Further studies on these factors, their interaction with each other and with their target genes, will be necessary, however, before the process of development is fully understood.

REFERENCES

Anderson, M. G., Perkins, G. L., Chittick, P., Shrigley, R. J. and Johnson, W.A. (1995). Drifter, a *Drosophilia* POU-domain transcription factor, is required for correct differentiation and migration of tracheal cells and midline glia. *Genes and Development* **9**, 123–127.

Biggin, M. D. and Tjian, R. (1989). A purified *Drosophila* homeodomain protein represses transcription *in vitro*. *Cell* **58**, 433–440.

Brickell, P. M. and Tickle, C. (1989). Morphogens in chick limb development. *BioEssays* **11**, 145–149.

Chan, S. K., Jaffe, L., Capovilla, M., Botas, J. and Mann, R. S. (1994). The DNA binding specificity of ultrabithorax is modulated by cooperative interactions with extradenticle, another homeoprotein. *Cell* **78**, 603–614.

de Kok, Y. J. M., Van der Maarel, S. M., Bitner-Glindzicz, M., Huber, I., Monaco, A. P., Malcolm, S., Pembrey, M. E., Ropers, H.-H. and Cremers, F. P. M. (1995). Association between X-linked mixed deafness and mutations in the POU domain gene POU3F4. *Science* **267**, 685–688.

Desai, C., Garriga, G., McIntire, S. L. and Horvitz, H. R. (1988). A genetic pathway for the development of *Caenorhabditis elegans* HSN motor neurons. *Nature* **336**, 638–646.

Driever, W. and Nusslein-Volhard, C. (1988a). A gradient of bicoid protein in *Drosophila* embryos. *Cell* **54**, 83–93.

Driever, W. and Nusslein-Volhard, C. (1988b). The bicoid protein determines position in the *Drosophila* embryo in a concentration-dependent manner. *Cell* **54**, 95–104.

Driever, W., Thoma, G. and Nusslein-Volhard, C. (1989). Determination of spatial domains of zygotic gene expression in the *Drosophila* embryo by the affinity of binding sites for the bicoid morphogen. *Nature* **340**, 363–367.

Edelman, G. M. and Jones, F. S. (1993). Outside and downstream of the homeobox. *Journal of Biological Chemistry* **268**, 20683–20686.

Evans, R. M. (1988). The steroid and thyroid hormone receptor gene super-family. *Science* **240**, 889–895.

Finney, M., Ruvkun, G. and Horvitz, H. R. (1988). The *C. elegans* cell lineage and differentiation gene *unc–86* encodes a protein with a homeodomain and extended similarity to transcription factors. *Cell* **55**, 757–769.

Gehring, W. J., Affoller, M. and Burglin, T. (1994). Homeodomain proteins. *Annual Review of Biochemistry* **63**, 487–526.

Graham, A., Papalopulu, N. and Krumlauf, R. (1989). The murine and *Drosophila* homeobox gene complexes have common features of organization and expression. *Cell* **57**, 367–378.

Hadorn, E. (1968). Transdetermination in cells. *Scientific American* **219**, 110–120.

Han, K., Levine, M.S. and Manley, J.L. (1989). Synergistic activation and repression of transcription by *Drosophila* homeobox proteins. *Cell* **56**, 573–583.

Harding, K., Weeden, C., McGinnis, W., and Levine, M. (1985). Spatially regulated expression of homeotic genes in *Drosophila*. *Science* **229**, 1236–1242.

Hayashi, S. and Scott, M. P. (1990). What determines the specificity of action of *Drosophila* homeodomain proteins. *Cell* **63**, 883–894.

He, X., Treacy, M. N., Simmons, D. M., Ingraham, H. A., Swanson, L. S. and Rosenfeld, M. G. (1989). Expression of a large family of POU-domain regulatory genes in mammalian brain development. *Nature* **340**, 35–42.

Hoey, T. and Levine, M. (1988). Divergent homeobox proteins recognize similar DNA sequences in *Drosophila*. *Nature* **332**, 858–861.

Holland, P. W. H. and Hogan, B. L. M. (1986). Phylogenetic distribution of Antennapedia-like homeoboxes. *Nature* **321**, 251–253.

Ingham, P. W. (1988). The molecular genetics of embryonic pattern formation in *Drosophila*. *Nature* **335**, 25–34.

Jaynes, J. B. and O'Farrell, P. H. (1988). Activation and repression of transcription by homeodomain-containing proteins that bind a common site. *Nature* **336**, 744–749.

Johnson, F. B. and Krasnow, M. A. (1992). Differential regulation of transcription preinitiation complex assembly by activator and repressor homeo domain protein. *Genes and Development* **6**, 2177–2189.

Kenyon, C. (1994). If birds can fly, why can't we? Homeotic genes and evolution. *Cell* **78**, 175–180.

Krasnow, M. A., Saffman, E. E., Kornfeld, K. and Hogness, D. S. (1989). Transcriptional activation and repression by Ultrabithorax proteins in cultured *Drosophila* cells. *Cell* **57**, 1031–1043.

Krumlauf, R. (1994). Hox genes in vertebrate development. *Cell* **78**, 191–201.

Lai, J. S., Cleary, M. A. and Herr, W. (1992). A single amino acid exchange transfers VP16-induced positive control from the Oct-1 to the Oct-2 homeo domain. *Genes and Development* **6**, 2058–2065.

Langston, A. W. and Gudas, L. S. (1994). Retinoic acid and homeobox gene regulation. *Current Opinion in Genetics and Development* **4**, 550–555.

Lawrence, P. A. and Morata, G. (1994). Homeobox genes: their function in *Drosophila* segmentation and pattern formation. *Cell* **78**, 181–189.

Marshall, H., Studer, M., Popperl, H., Aparicio, S., Kurolwa, A., Brenner, S. and Krumlauf, R. (1994). A conserved retinoic acid response element required for early expression of the homeobox gene Hoxb-1. *Nature* **370**, 567–571.

Pignoni, F., Baldarelli, R. M., Steingrimsson, E., Diaz, R. J., Patapoutian, A., Merriam, J. R. and Lengyel, J. A. (1990). The *Drosophila* gene *tailless* is expressed at the embryonic termini and is a member of the steroid receptor superfamily. *Cell* **62**, 151–163.

Rosenberg, U. B., Schroder, C., Preiss, A., Kienlin, A., Cote, S., Riede, I. and Jackle, H. (1986). Structural homology of the *Drosophila Kruppel* gene with *Xenopus* transcription factor III A. *Nature* **319**, 336–339.

Small, S., Krant, R., Hoey, T., Warrior, R. and Levine, M. (1991).

Transcriptional regulation of a pair-rule stripe in *Drosophila. Genes and Development* **5**, 827–839.

Struhl, G., Johnston, P. and Lawrence, P.A. (1992). Control of *Drosophila* body pattern by the hunchback morphogen gradient. *Cell* **69**, 237–249.

Tautz, D. (1988). Regulation of the *Drosophila* segmentation gene hunchback by two maternal morphogenetic centres. *Nature* **332**, 281–284.

Van Dijk, M. A. and Murre, C. (1994). Extradenticle raises the DNA binding specificity of homeotic selector gene products. *Cell* **78**, 617–624.

Verrijzer, C. R. and van der Vliet, P. C. (1993). POU domain transcription factors. *Biochimica et Biophysica Acta* **1173**, 1–21.

Verrijzer, C. R., Kal, A. J. and van der Vliet, P. C. (1990). The DNA binding domain (POU domain) of transcription factor Oct-1 suffices for stimulation of DNA replication. *EMBO Journal* **9**, 1883–1888.

Voss, J. W. and Rosenfeld, M. G. (1992). Anterior pituitary development: short tales from dwarf mice. *Cell* **70**, 527–530.

Way, J. C. and Chalfie, M. (1988). *mec-3*, a homeobox-containing gene that specifies differentiation of the touch receptor neurons in *C. elegans. Cell* **54**, 5–16.

Wegner, M., Drolet, D. W. and Rosenfeld, M. G. (1993). POU domain proteins: structure and function of development regulators. *Current Opinion in Cell Biology* **5**, 488–498.

Weigel, D. and Meyerowitz, E. M. (1994). The ABCs of floral homeotic genes. *Cell* **78**, 203–209.

Winslow, G. M., Hayashi, S., Krasnow, M., Hogness, D. S. and Scott, M.P. (1989). Transcriptional activation and repression by the *Antennapedia* and *fushi tarazu* proteins in cultured *Drosophila* cells. *Cell* **57**, 1017–1030.

CHAPTER SEVEN

Transcription factors and cancer

7.1 INTRODUCTION

In previous chapters we have discussed the involvement of transcription factors in normal cellular regulatory processes, for example constitutive, inducible, cell type-specific or developmentally regulated transcription. It is not surprising that aspects of this complex process can go wrong and that the resulting defects in transcription factors can result in disease. For example, the absence of a specific transcription factor necessary for the transcription of the HLA class II genes results in the lack of transcription of these genes and the absence of their protein products, which is characteristic of one form of congenital severe combined immunodeficiency disease (Reith *et al.*, 1988). Similarly, as noted in Section 6.3.2, mutations in the gene encoding the human Pit–1 transcription factor result in a failure of pituitary gland development, leading to congenital dwarfism.

A special place in the human diseases that can involve alterations in transcription factors is occupied, however, by cancer. Thus, because this disease results from growth in an inappropriate place or at an inappropriate time, it can be caused not only by deficiencies in particular genes but also by the enhanced expression or activation of specific cellular genes involved in growth regulatory processes which are normally expressed only at low levels or very transiently.

Interestingly, many cancer-causing genes of this type, known as oncogenes (reviews: Bishop, 1987; Bourne and Varmus, 1992), were originally identified within cancer-causing retroviruses which had picked them up from the cellular genome. Within the virus, the onco-gene has become activated either by over-expression or by mutation and is therefore responsible for the ability of the virus to transform cells to a cancerous phenotype. In contrast, the homologous gene within the cellular genome is clearly not always cancer-causing, since all cells are not cancerous. It can be activated, however, into a cancer-causing form either by over-expression or by mutation, and hence these genes can play an important role in the generation of human cancer (Figure 7.1). The forms of the oncogene isolated from the retrovirus and from the normal cellular genome are distinguished by the prefixes v and c respectively, as in v-*onc* and c-*onc*.

Despite this potential to cause cancer, the c-*onc* genes are highly conserved in evolution, being found not only in the species from which the original virus was isolated but in a wide range of other eukaryotes. This indicates that the products of these oncogenes play a critical role in the regulation of normal cellular growth processes, their malregulation or mutation, resulting, therefore, in abnormal growth and cancer. In agreement with this idea, oncogenes identified in this way include genes encoding many different types of protein involved in growth control, such as growth factors, growth factor receptors and G proteins. They also include, however, several genes encoding cellular transcription

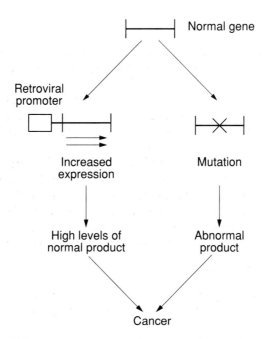

Figure 7.1 A cellular proto-oncogene can be converted into a cancer-causing oncogene by increased expression or by mutation.

factors that normally regulate specific sets of target genes. Similarly, a number of other genes encoding transcription factors have been identified at the break-points of the chromosomal translocations characteristic of human leukaemias, with their activation being involved in the resulting cancer. Section 7.2 therefore discusses several cases of this type and the insights they have provided into the processes regulating gene expression in normal cells and their malregulation in cancer (reviews: Gutman and Wasylyk, 1991; Forrest and Curran, 1992; Rabbits, 1994).

Following the discovery of cellular oncogenes, it subsequently became clear that another class of genes existed whose protein products appeared to restrain cellular growth. The deletion or mutational inactivation of these so-called anti-oncogenes therefore results in the abnormal unregulated growth characteristic of cancer cells (reviews: Knudson, 1993; Weinberg, 1993). As some of these anti-oncogenes also encode transcription factors, they are discussed in Section 7.3.

7.2 CELLULAR ONCOGENES AND CANCER

7.2.1 Fos, Jun and AP1

As noted in Table 4.1, the AP1-binding site is a DNA sequence that renders genes which contain it inducible by treatment with phorbol esters such as TPA. The activity binding to this site is referred to as AP1 (activator protein 1). It is clear, however, that preparations of AP1 purified by affinity chromatography on an AP1-binding site contain several different proteins (reviews: Curran and Franza, 1988; Ransone and Verma, 1990; Kerppola and Curran, 1995).

A possible clue as to the identity of one of these AP1-binding proteins was provided by the finding that the yeast protein GCN4, which induces transcription of several yeast genes involved in amino acid biosynthesis, does so by binding to a site very similar to the AP1 site (Figure 7.2). In turn, the DNA-binding region of GCN4 shows strong homology at the amino acid level to v-*jun*, the oncogene of avian sarcoma virus ASV17

DNA-binding site

GCN4	5'	T	G	A	C/G	T	C	A	T 3'
AP1	5'	T	G	A	G	T	C	A	G 3'

Figure 7.2 Relationship of the DNA-binding sites for the yeast transcription factor GCN4 and the mammalian transcription factor AP1.

Jun   206 P L F P I D M E S Q E R I K A E R K R M R N R I A A S K S R K
GCN4 216 P L S P I V P E S S D P   A A L K R A R N T E A A R R S R A

R K L E R I A R L E E K V K T L K A Q N S E L A S T A N M L R
R K L Q R M K Q L E D K V   E E L L S K N Y H L E

E Q V A Q L K Q K V M N H V N S G C Q L M L T Q Q L G T F 296
N E V A R L K K L V G E R 281

Figure 7.3 Comparison of the C-terminal amino acid sequences of the chicken Jun protein and the yeast transcription factor GCN4. Boxes indicate identical residues.

(Figure 7.3; Vogt *et al.*, 1987). This suggested, therefore, that the protein encoded by the cellular homologue of this gene, c-*jun*, which was known to be a nuclearly located DNA-binding protein, might be one of the proteins that bind to the AP1 site.

In agreement with this, antibodies against the Jun protein react with purified AP1-binding proteins, whilst Jun protein expressed in bacteria can bind to AP1-binding sites (Angel *et al.*, 1988). Hence the Jun protein is capable of binding to the AP1-binding site and constitutes one component of purified AP1 preparations, which also contain other Jun-related proteins such as Jun B (Figure 7.4). Moreover, co-transfection of a vector expressing the Jun protein with a target promoter resulted in increased transcription if the target gene contained AP1-binding sites but not if it lacked them, indicating that Jun was capable of stimulating transcription via the AP1 site (Figure 7.5, Angel *et al.*, 1988). Hence the Jun oncogene product is a sequence-specific transcription factor capable of stimulating transcription of genes containing its binding site.

Figure 7.4 Passage of total cellular proteins through a column containing an AP1 site results in the purification of several cellular proteins, including Jun, Jun B, Fos and Fos-related antigens (Fras), which are capable of binding to the AP1 site either alone or in combination.

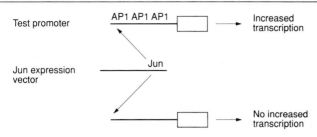

Figure 7.5 Artificial expression of the Jun protein in an expression vector results in the activation of a target promoter containing several AP1-binding sites but has no effect on a similar promoter lacking these sites, indicating that Jun can specifically activate gene expression via AP1-binding sites.

In addition to Jun and Jun-related proteins, purified AP1 preparations also contain the product of another oncogene, c-*fos*, as well as several Fos-related proteins known as the Fras (Fos-related antigens; see Figure 7.4). Unlike Jun, however, Fos cannot bind to the AP1 site alone but can do so only in the presence of another protein, p39, which is identical to Jun (Rauscher *et al.*, 1988). Hence, in addition to its ability to bind to AP1 sites alone, Jun can also mediate binding to this site by the Fos protein. Such DNA binding by Fos and Jun is dependent on the formation of a dimeric molecule. Although Jun can form a DNA-binding homodimer, Fos cannot do so. Hence DNA binding by Fos is dependent upon the formation of a heterodimer between Fos and Jun, which binds to the AP1 site with approximately 30-fold greater affinity than the Jun homodimer (Figure 7.6; Halazonetis *et al.*, 1988).

It is clear, therefore, that both Fos and Jun, which were originally isolated in oncogenic retroviruses, are also cellular transcription factors which play an important role in inducing specific cellular genes following phorbol ester treatment (review: Ransone and Verma, 1990). Increased levels of Fos and Jun occur in cells following treatment with phorbol esters (Lamph *et al.*, 1988), indicating that these substances act,

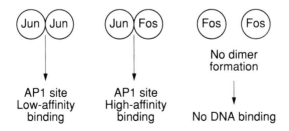

Figure 7.6 Heterodimer formation between Fos and Jun results in a complex capable of binding to an AP1 site with approximately 30-fold greater affinity than a Jun homodimer, whilst a Fos homodimer cannot bind to the AP1 site.

at least in part, by increasing the levels of Fos and Jun, which in turn bind to the AP1 sites in phorbol ester-responsive genes and activate their expression.

Similar increases in the levels of Fos and Jun, as well as of Jun-B and the Fos-related protein Fra–1, are also observed when quiescent cells are stimulated to grow by treatment with growth factors or serum (Lamph *et al.*, 1988; Rauscher *et al.*, 1988), indicating that these substances act, at least in part, by increasing the levels of Fos and Jun, which in turn will switch on genes whose products are necessary for growth itself (Figure 7.7). In agreement with this idea, cells derived from mice in which the c-*jun* gene has been inactivated grow very slowly in culture and the mice themselves die early in embryonic development (Johnson *et al.*, 1993). Hence Fos and Jun play a critical role in normal cells, as transcription factors inducing phorbol ester or growth-dependent genes.

Normally, levels of Fos and Jun increase only transiently following growth factor treatment, resulting in a period of brief controlled growth. Clearly, continuous increase in the levels of these proteins, such as would occur when cells become infected with a retrovirus expressing one of them, would result in cells that exhibited continuous uncontrolled

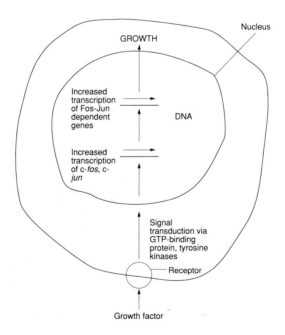

Figure 7.7 Growth factor stimulation of cells results in increased transcription of the c-*fos* and c-*jun* genes, which in turn stimulates transcription of genes that are activated by the Fos–Jun complex.

growth and were not subject to normal growth regulatory signals. Since such uncontrolled growth is one of the characteristics of cancer cells, it is relatively easy to link the role of Fos and Jun in inducing genes required for growth with their ability to cause cancer. Normally, however, the transformation of a cell to a transformed cancerous phenotype requires more than simply its conversion to a continuously growing immortal cell (review: Land *et al.*, 1983). Since repeated treatments with phorbol esters can promote tumour formation in immortalized cells, the prolonged induction of phorbol ester-responsive genes by raised levels of Fos and Jun may therefore result in the conversion of already continuously growing cells into the tumorgenic phenotype characteristic of cancer cells (Figure 7.8).

Hence the ability of Fos and Jun to cause cancer represents an aspect of their ability to induce transcription of specific cellular genes. In agreement with this idea, mutations in Fos that abolish its ability to

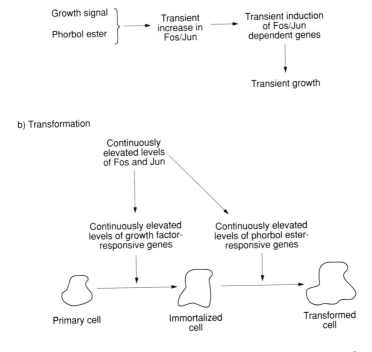

Figure 7.8 Effects of Fos and Jun on cellular growth. (a) In normal cells a brief exposure to a growth signal or phorbol ester will lead to a brief period of growth via the transient induction of Fos and Jun and hence of Fos/Jun-dependent genes. (b) In contrast, the continuous increase in the levels of Fos and Jun, produced, for example, by infection with a retrovirus expressing Fos or Jun, results in continuous unlimited growth and cellular transformation.

dimerize with Jun and hence prevent it from binding to AP1 sites also abolish its ability to transform cells to a cancerous phenotype (Schuermann *et al.*, 1989). It should be noted, however, that, in addition to their over-expression within a retrovirus, there is also some evidence that mutational changes render the viral proteins more potent transcriptional activators than the equivalent cellular proteins. Thus the v-Jun protein appears to activate transcription more efficiently than c-Jun due to a deletion in a region involved in targeting the c-Jun protein for degradation (Treier *et al.*, 1994) and which also mediates its interaction with a negatively acting cellular factor (Baichwal and Tjian, 1990).

Interestingly, in addition to its central role in the growth response, the Fos–Jun–AP1 system also appears to represent a target for other oncogenes. Thus, for example, the *ets* oncogene which, like *fos* and *jun*, encodes a cellular transcription factor, acts via a DNA-binding site, known as PEA3, which is located adjacent to the AP1 site in a number of TPA-responsive genes such as collagenase and stromelysin, and the Ets protein co-operates with Fos and Jun to produce high-level activation of these promoters (Wasylyk *et al.*, 1990). In addition to interacting positively with other factors, the Fos–Jun complex can also inhibit the action of other transcription factors. Thus, as described in Section 4.3.4, the Fos–Jun complex can interact with the activated glucocorticoid receptor to prevent it binding to DNA and activating transcription. Similarly, both Fos and Jun can inhibit the activation of muscle-specific promoters by the MyoD transcription factor (see Section 5.3), thereby preventing cells from differentiating into non-dividing muscle cells and allowing cellular proliferation to continue (Li *et al.*, 1992).

Hence the Fos and Jun oncogene products play a critical role in the regulation of specific cellular genes in normal cells, interacting with the products of other transcription factors to produce the controlled activity of their target genes necessary for normal controlled growth.

7.2.2 v-*erbA* and the thyroid hormone receptor

The v-*erbA* oncogene is one of two oncogenes carried by avian erythroblastosis virus (AEV). The cellular equivalent of this oncogene, c-*erbA*, has been shown to encode the cellular receptor for thyroid hormone (Sap *et al.*, 1986; Weinberger *et al.*, 1986), which is a member of the steroid–thyroid hormone receptor superfamily discussed in Section 4.3. Following the binding of thyroid hormone, the receptor–hormone complex binds to its appropriate recognition site in the DNA of thyroid-responsive genes and activates their transcription (Figure 7.9).

Hence the protein encoded by the c-*erbA* gene represents a bona fide cellular transcription factor involved in the activation of thyroid-respon-

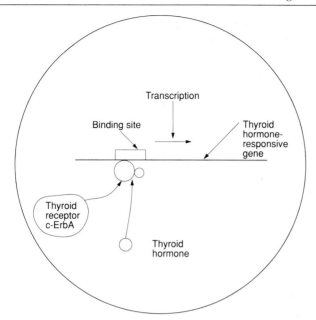

Figure 7.9 The c-*erb*A gene encodes the thyroid hormone receptor and activates transcription in response to thyroid hormone.

sive genes. Unlike the case of the *fos* and *jun* gene products, which regulate genes involved in growth, it is not immediately obvious how the form of thyroid hormone receptor encoded by the viral v-*erb*A gene can transform cells to a cancerous phenotype.

The solution to this problem is provided by a comparison of the cellular ErbA protein, which is a functional thyroid hormone receptor, and the viral ErbA protein encoded by AEV. Thus, in addition to being fused to the retroviral gag protein at its N-terminus, the viral ErbA protein contains several mutations in the regions of the receptor responsible for binding to DNA and for binding thyroid hormone, as well as a small deletion in the hormone-binding domain (Figure 7.10).

Interestingly, although these changes do not abolish the ability of the viral ErbA protein to bind to DNA, they do prevent it from binding thyroid hormone and thereby becoming converted to a form that can activate transcription (Sap *et al.*, 1986; 1989). However, these changes do not affect the inhibitory domain which, as discussed in Section 4.3.4, allows the thyroid hormone receptor to repress transcription. Hence, the viral v-ErbA protein can inhibit the induction of thyroid hormone-responsive genes when cells are treated with thyroid hormone by binding to the thyroid hormone response elements in their promoters and dominantly repressing their transcription, as well as preventing binding

Figure 7.10 Relationship of the cellular ErbA protein and the viral protein. The black dots indicate single amino acid differences between the two proteins, while the arrow indicates the region where nine amino acids are deleted in the viral protein.

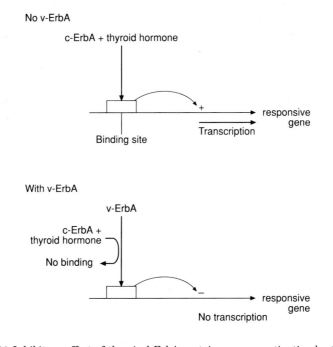

Figure 7.11 Inhibitory effect of the viral ErbA protein on gene activation by the cellular protein, in response to thyroid hormone. The viral protein inhibits binding by the activated c-ErbA protein and also dominantly represses transcription by means of its inhibitory domain. Note the similarity to the action of the α2 form of the c-ErbA protein, illustrated in Figure 4.20.

of the activating complex of thyroid hormone and the cellular ErbA protein (Figure 7.11).

Hence the viral ErbA protein acts as a dominant repressor of thyroid-responsive genes, being both incapable of activating transcription itself and able to prevent activation by intact receptor. This mechanism of action is clearly similar to the repression of thyroid hormone-responsive genes by the naturally occurring alternatively spliced form of the thyroid hormone receptor which, as discussed in Section 4.3.4, lacks the hormone-binding domain and therefore cannot bind hormone. Thus the same mechanism of gene repression by a non-hormone-binding receptor is used naturally in the cell and by an oncogenic virus.

One of the targets for repression by the viral ErbA protein is the erythrocyte anion transporter gene (Zenke *et al.*, 1988), which is one of the genes normally induced when avian erythroblasts differentiate into erythrocytes. This differentiation process has been known for some time to be inhibited by the ErbA protein and it is now clear that it achieves this effect by blocking the induction of the genes needed for differentiation. In turn such inhibition will allow continued proliferation of these cells, rendering them susceptible to transformation into a tumour cell type by the product of the other AEV oncogene, v-*erbB*, which encodes a truncated form of the epidermal growth factor receptor (Downward *et al.*, 1984) and therefore renders cell growth independent of external growth factors (Figure 7.12).

The two cases of Fos–Jun and ErbA therefore represent contrasting examples of the involvement of transcription factors in oncogenesis, in

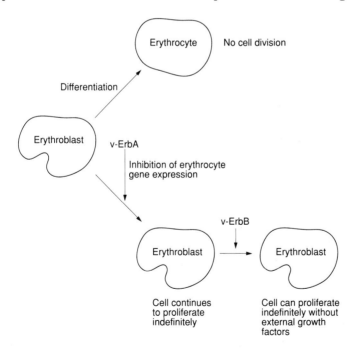

Figure 7.12 Inhibition of erythrocyte-specific gene expression by the v-ErbA protein prevents erythrocyte differentiation and allows transformation by the v-ErbB protein.

terms of both the mechanism of transformation and the manner in which the cellular form of the oncogene becomes an active transforming gene. Thus, in the case of Fos and Jun, transformation is achieved by the continuous activation of genes necessary for growth in normal cell types. Moreover, it occurs, at least in part, via the natural activity of the cellular oncogene in inducing these genes being enhanced by their over-expression such that it occurs at an inappropriate time or place (Figure 7.13a). In contrast, in the ErbA case, transformation is achieved by inhibiting the expression of genes whose products are required for the differentiation of a particular cell type, therefore allowing growth to continue. Moreover, this occurs via the activity of a mutated form of the transcription factor which, rather than carrying out its normal function more efficiently, actually interferes with the normal role of the thyroid hormone receptor in inducing thyroid hormone-responsive genes required for differentiation (Figure 7.13b).

7.2.3 The *myb* oncogene

In the case of the *fos/jun* and *erbA* oncogenes it was relatively easy to identify the binding sites within DNA and the target genes whose expression is affected by the corresponding cellular transcription factors. This was not the case, however, in the case of the *myb* oncogene, originally isolated from two avian retroviruses E26 and AMV (reviews: Luscher and Eisenman, 1990; Graf, 1992). Thus, although both v-Myb and c-Myb proteins are located in the nucleus and bind to DNA, they did not appear to be related to any known transcription factor or DNA-binding activity.

To identify the binding site of the Myb protein, therefore, Biedenkapp *et al.* (1988) screened random clones of chicken DNA with purified Myb protein in order to identify DNA sequences that could bind Myb protein. They identified six of 1200 clones which did so and showed that each of these clones contained multiple copies of the sequence PyAACC/TG, over which the Myb protein formed a DNAse I footprint (Section 2.2.2), allowing the Myb-binding site to be identified (Figure 7.14). Subsequently it was shown that a plasmid expressing the Myb protein can transactivate a target promoter containing its binding site, indicating that the Myb protein is a transcriptional activator which functions by binding to a specific site in the DNA (Weston and Bishop, 1989).

Although such studies clearly identify the Myb-binding site and demonstrate its ability to activate transcription, they do not identify the actual cellular genes that are normally activated by Myb. To do this Ness *et al.* (1989) made use of a temperature-sensitive form of the Myb protein which transforms cells at low temperatures but is inactive and

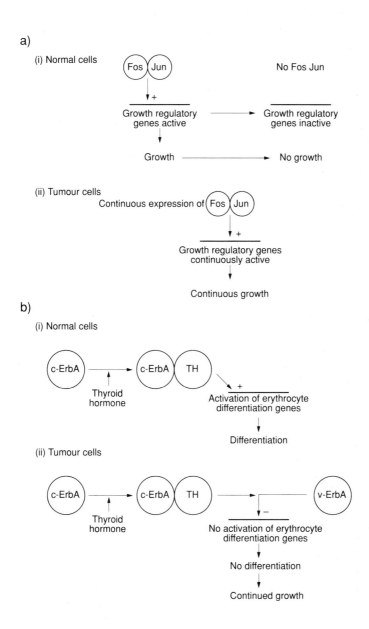

Figure 7.13 Transformation mechanisms of Fos/Jun (a) and ErbA (b). Note that Fos/Jun-induced transformation occurs because the proteins induce the continual activation of growth regulatory genes which are normally expressed only transiently, whilst v-ErbA-induced transformation occurs because the protein interferes with the action of its cellular homologue and hence inhibits the induction of genes involved in erythrocyte differentiation.

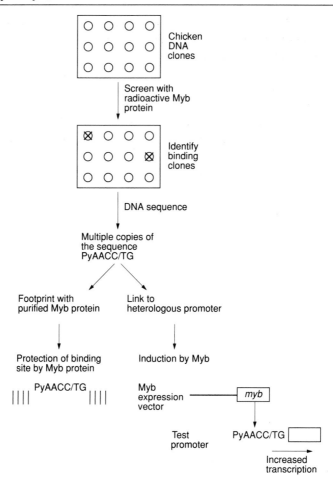

Figure 7.14 Identification of the DNA-binding site recognized by the Myb protein. A large number of clones containing random pieces of chicken DNA were screened for the ability to bind radioactive Myb proteins. Clones that bound the protein were subsequently all shown to possess multiple copies of a specific DNA sequence, indicating that this was the Myb-binding site. This was subsequently confirmed by showing that purified Myb protein would produce a DNAse I footprint on this DNA (see Figure 2.4) and that Myb protein produced in cultured cells from an expression vector could activate a target gene containing this sequence in its promoter (compare Figure 7.5).

fails to transform at higher temperatures. Hence, in cells transformed with a virus encoding this protein, the protein can be inactivated and the transformed state reversed by raising the temperature. By using the technique of differential cDNA library screening (which was also used to isolate the MyoD gene—see Section 5.3.1), Ness *et al.* isolated a gene, *mim*-1, which was expressed in the transformed cells only at low temperature (37°C) when the Myb protein was active and not at high temperature (42°C) when it was inactive (Figure 7.15). The *mim*-1 gene was also expressed at high levels in cells transformed with the normal Myb protein but not in untransformed cells.

The conclusion from these experiments, that *mim*-1 is a gene activated by Myb, was confirmed by the finding that the *mim*-1 promoter contains three consensus binding sites for the Myb protein and that this promoter can be activated by co-transfection with a plasmid expressing the Myb protein (Ness *et al.*, 1989). Hence Myb is clearly a sequence-

Figure 7.15 Identification of the *mim*-1 gene as a gene whose expression is induced by Myb expression. Cells were infected with a retrovirus expressing a temperature-sensitive form of the Myb protein which was active at 37°C but inactivated when the temperature was raised to 42°C. Subtractive hybridization techniques were then used to isolate a gene whose corresponding mRNA was present in cells at 37°C which were expressing active Myb but not in cells maintained at 42°C and hence expressing inactive Myb. The single gene isolated in this way was named *mim*-1 (compare the use of a similar technique to isolate MyoD; see Figure 5.8).

specific transcription factor which activates the transcription of at least one cellular gene.

Both c-*myb* and the *mim*-1 gene are expressed at high levels in early differentiation stages of the monocyte lineage, with levels of expression declining as these cells differentiate towards mature macrophages (Ness *et al.*, 1989). The switching on of Myb expression in the early stages of monocyte differentiation plays a key role in the differentiation process. Thus the ectopic expression of Myb together with the NF-M transcription factor (a member of the C/EBP family—see Section 3.3.3) in cells of a completely different type, such as fibroblasts or erythroid cells, results in the activation of genes characteristic of monocytes such as *mim*-1 and lysozyme (Ness *et al.*, 1993) (Figure 7.16). Conversely, the subsequent decline in c-Myb expression in the monocyte lineage results in a corresponding fall of *mim*-1 expression and differentiation of monocytes to mature non-dividing macrophages. Interestingly, the v-*myb* oncogene specifically transforms the early monocytic differentiation stages, and can actually reverse the differentiation state of relatively mature macrophages, producing more immature cells (Ness *et al.*, 1987).

Hence, like the ErbA protein, the Myb protein transforms cells by

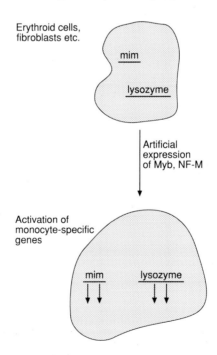

Figure 7.16 The artifical expression of Myb and the C/EBP family transcription factor NF-M in non-monocyte cell types is sufficient to activate the expression of monocyte-specific genes such as *mim*-1 and lysozyme.

inhibiting or reversing their differentiation, allowing continued proliferation to occur. This is apparently achieved, however, by activating genes required in the undifferentiated cells, rather than, as in the case of ErbA, by repressing genes whose products are required for differentiation (Figure 7.17). Thus, as discussed above, the Myb protein is a potent trans-activator, capable of inducing the transcription of genes containing its binding site, including the *mim*-1 gene. Moreover, the ability of various mutant Myb proteins to transform cells correlates precisely with their ability to transactivate gene expression (Lane *et al.*, 1990), indicating that this ability is essential for transformation.

Interestingly, as in the case of c-*erbA* and v-*erbA*, the products of the c-*myb* and v-*myb* genes differ from one another. In the case of the *myb* genes, however, the v-*myb* gene lacks the region encoding the N-terminal region of the protein. This region contains a site for phosphorylation, such phosphorylation inhibiting the DNA-binding activity of the intact cellular protein. Hence, in contrast to the cellular Myb protein whose DNA-binding activity is regulated by phosphorylation and dephosphorylation, the viral Myb protein has constitutive DNA-binding activity and is therefore more readily able to activate its target genes (Luscher *et al.*, 1990; Figure 7.18). Hence the *myb* and *erbA* cases represent two contrasting examples of the creation of a transforming protein by mutation. Thus the viral Myb protein has been altered by mutation into a more potent transcriptional activator capable of inducing genes required in undifferentiated cell types, whilst the ErbA protein has been altered into a repressor capable of inhibiting the

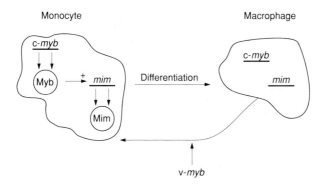

Figure 7.17 Differentiation of cells from immature monocytes to mature macrophages is associated with a switching off of c-*myb* expression and hence of *mim*-1. Artificial expression of v-*myb* in macrophages is capable of reversing their differentiated phenotype and converting them to monocytes, presumably by activating the expression of *mim*-1 and/or other genes expressed in undifferentiated monocytes but not in mature macrophages.

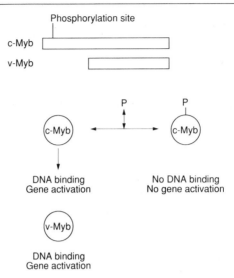

Figure 7.18 Comparison of the c-Myb and v-Myb proteins. Note that v-Myb lacks the N-terminal portion of the protein which contains a phosphorylation site. DNA-binding ability of the c-Myb protein can be inhibited by phosphorylation of this site, allowing its activity to be regulated. In contrast, since v-Myb lacks this site its activity cannot be controlled in this way.

activating effects of its cellular homologue on genes required in differentiated cells.

As well as presenting a contrasting case to that of the ErbA protein, the studies on the Myb protein discussed in this section represent an excellent example of the power of molecular techniques in characterizing the properties of a factor that was originally identified solely on the basis of its presence in a transforming virus and which is now believed to play a critical role in the growth and differentiation of monocytic cells and in their transformation.

7.2.4 The *myc* oncogene

Interestingly, for a considerable period, the techniques of molecular biology failed in the case of the c-*myc* oncogene, which was one of the earliest cellular oncogenes to be identified, its expression being dramatically increased in a wide variety of transformed cells (review: Spencer and Groudine, 1991). Thus the Myc protein has a number of properties suggesting that it is a transcription factor, notably nuclear localization, the possession of several motifs characteristic of transcription factors such as the helix–loop–helix and leucine zipper elements (see Section

Figure 7.19 Both Myc–Max heterodimers and Max–Max homodimers can bind to DNA, whereas Myc–Myc homodimers cannot.

8.4), and the ability to activate target promoters in co-transfection assays. Despite exhaustive efforts, however, no DNA sequence to which the Myc protein binds could be defined, rendering its mechanism of action uncertain.

The solution to this problem was provided by the work of Blackwood and Eisenman (1991), who identified a novel protein, Max, which can form heterodimers with the Myc protein via the helix–loop–helix motif present in both proteins. Myc–Max heterodimers can bind to DNA and regulate transcription, whereas Myc–Myc homodimers cannot (reviews: Blackwood *et al.*, 1992; Amati and Land, 1994) (Figure 7.19). This effect evidently parallels the requirement of the Fos protein for dimerization with Jun in order to bind with high affinity to AP1 sites (see Section 7.2.1).

The Max protein therefore plays a critical role in allowing the DNA binding of Myc. Moreover, the ability to interact with Max, bind to DNA and modulate gene expression is critical for the ability of the Myc protein to transform, since mutations in Myc that abolish its ability to heterodimerize with Max also abolish its transforming ability (Amati *et al.*, 1993). Hence, as was previously speculated, the Myc protein is a transcription factor whose over-expression causes transformation, presumably via the activation of genes whose protein products are required for cellular growth.

Interestingly, the Max protein does not appear to represent a passive partner which merely serves to deliver Myc to the DNA of target genes. Rather, it plays a key role in regulating the activity of target genes containing the appropriate binding site. Thus, it has been shown that, whereas Myc–Max heterodimers can activate transcription, Max–Max homodimers can bind to the same site and repress transcription (Kretzner *et al.*, 1992). Interestingly, Myc is expressed at very low levels in resting cells and its expression is induced when cells begin to grow, whereas Max is expressed at similar high levels in both resting and proliferating cells. Hence, in resting cells, the expression of Myc-dependent genes will be repressed by Max–Max homodimers whilst expression will be activated by Myc–Max heterodimers as the cells receive

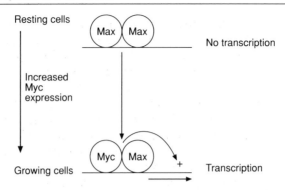

Figure 7.20 In resting cells, Myc-dependent genes will be repressed by a Max–Max homodimer. As cells begin to grow, the expression of Myc increases, resulting in the formation of Myc–Max homodimers which activate transcription.

signals to proliferate, resulting in increased Myc expression (Figure 7.20). Clearly the over-expression of the Myc gene, which is observed in many cancer cells, would result in a similar production of activating Myc–Max heterodimers leading to gene activation. Hence, as in the case of the Fos–Jun system, transformation by the Myc oncogene appears to depend primarily on its over-expression, resulting in the activation of genes required for cellular growth.

7.2.5 Other oncogenic transcription factors

In view of the likely need for multiple different transcription factors to regulate genes involved in cellular growth processes, it is not surprising that several other genes encoding transcription factors have also been identified as oncogenes. Thus, for example, the *rel* oncogene of the avian retrovirus Rev-T is a member of the NFκB family of transcription factors discussed in Section 5.2.2 (review: Gilmore, 1990), whilst the *Bcl*-3 oncogene is a member of the IκB family which interacts with the NFκB proteins (Bours *et al.*, 1993). Interestingly, the Bcl-3 factor illustrates another facet of the mechanisms by which transcription factor genes become oncogenic. Thus this factor was not identified as a retroviral oncogene but on the basis that it was located at the break-point of chromosomal rearrangements, which resulted in its translocation to a position adjacent to the immunoglobulin gene in some B cell chronic leukaemias.

A number of other transcription factors have also been shown to be capable of causing cancer when translocated in this way, either because their expression is increased or because of the production of a novel form of the protein due to its truncation or linkage to other genes following the translocation. These include both factors that were originally identified in oncogenic retroviruses and others that had not previously been shown to have oncogenic potential (review: Rabbits, 1994). Thus, for example, expression of the c-*myc* oncogene (Section 7.2.4) is dramatically increased by its translocation into the immunoglobulin heavy-chain locus, which occurs in the human B cell malignancy known as Burkitt's lymphoma (review: Spencer and Groudine, 1991), whilst the Ets transcription factor gene discussed above (Section 7.2.1) is fused to the gene for platelet-derived growth factor receptor to create a novel oncogenic fusion protein in patients with chronic myelomonocytic leukaemia (review: Sawyers and Denny, 1994). Similarly, expression of the homeobox gene Hox 11 (see Section 6.3) is activated in cases of acute childhood T cell leukaemia, whilst the retinoic acid receptor α (see Section 4.3.4) is fused to the PML gene (which itself encodes a zinc finger-containing transcription factor) in acute promyelocytic leukaemia (Dyck *et al.*, 1994).

These findings provide further evidence that transcription factor genes are not only rendered oncogenic by transfer into a retrovirus but are also involved in the causation of human cancers, playing a key role, for example, in the oncogenic effects of the chromosome translocations that are characteristic of specific cancers.

7.3 ANTI-ONCOGENES AND CANCER

7.3.1 Nature of anti-oncogenes

As noted in Section 7.1, a number of genes exist whose normal function is to encode proteins that function in an opposite manner to those of oncogenes, acting to restrain cellular growth. The deletion or mutational inactivation of these anti-oncogenes (also known as tumour suppressor genes) therefore results in cancer (reviews: Knudson, 1993; Weinberg, 1993) (Figure 7.21). This effect evidently parallels the production of cancer by the over-expression or mutational activation of cellular proto-oncogenes (compare Figures 7.1 and 7.21).

A number of anti-oncogenes of this type have been defined, and three encode transcription factors. Two of these, p53 and the Wilms' tumour gene product, act by binding to the DNA of their target genes and regulating their expression, whereas the retinoblastoma gene product (Rb-1) acts primarily via protein–protein interactions with other DNA-binding transcription factors. The p53 and Rb-1 proteins are therefore discussed in Sections 7.3.2 and 7.3.3 as examples of these two mechan-

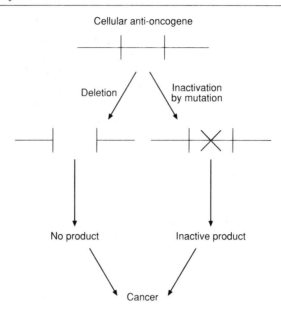

Cellular anti-oncogene

Deletion

Inactivation
by mutation

No product

Inactive product

Cancer

Figure 7.21 Cancer can result from the deletion of specific anti-oncogenes or their inactivation by mutation.

isms of action. For discussion of the Wilms' tumour gene product, see Hastie (1993).

7.3.2 p53

The gene encoding the 53-kDa protein known as p53 is mutated in a wide variety of human tumours, especially carcinomas (review: Levine *et al.*, 1991). In normal cells, expression of this protein is induced by agents that cause DNA damage, and its over-expression results in growth arrest of cells containing such damage or their death by the process of programmed cell death (apoptosis). Hence p53 has been called the 'guardian of the genome' (Lane, 1992), which allows cells to proliferate only if they have intact undamaged DNA. This would prevent the development of tumours containing cells with mutations in their DNA, and inactivation of the p53 gene by mutation would therefore result in an enhanced rate of tumour formation. In agreement with this idea, mice in which the p53 gene has been inactivated do not show any abnormalities in normal development but do exhibit a high rate of tumour formation, leading to early death (review: Berns, 1994).

The molecular analysis of the p53 gene product (review: Vogelstein and Kinzler, 1992) showed that it contains a DNA-binding domain and a region capable of activating transcription. The majority of the mutations in p53 that occur in human tumours are located in the DNA-binding domain (Friend, 1994; Anderson and Tegtmeyer, 1995). These mutations result in a failure of the mutant p53 protein to bind to DNA, indicating that this ability is crucial for the ability of the normal p53 protein to regulate cellular growth and suppress cancer.

The p53 protein therefore functions, at least in part, by activating the

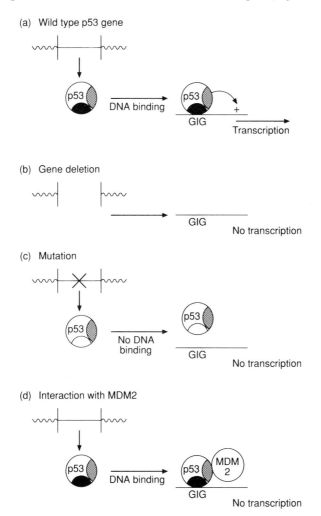

Figure 7.22 The ability of wild-type p53 to activate genes encoding growth inhibiting proteins (GIGs) (a) can be abolished by deletion of the p53 gene (b), by mutations in the DNA-binding domain (solid) which prevent it binding to DNA (c) or by the interaction of functional p53 with the MDM2 protein, which masks its transcriptional activation domain (hatched) thereby preventing it from activating transcription (d).

expression of genes whose protein products act to inhibit cellular growth (Figure 7.22a). The absence of functional p53, due either to gene deletion (Figure 7.22b) or to its inactivation by mutation (Figure 7.22c), results in a failure to express these genes, leading to uncontrolled growth. In addition, functional p53 can also be prevented from activating gene transcription by interaction with the MDM2 oncoprotein, which masks the transcriptional domain of p53 (Oliner *et al.*, 1993) (Figure 7.22d). This effect is of particular importance in many human soft tissue sarcomas where the p53 gene is intact and encodes wild-type p53, but the protein is functionally inactivated due to the high levels of MDM2 resulting from amplification of the *mdm2* gene encoding it.

A similar interaction of p53 with oncogenic proteins also occurs with the transforming proteins of several DNA viruses. Indeed, p53 was originally discovered as a protein that interacted with the large T oncoprotein of the DNA tumour virus SV40. The functional inactivation of p53 produced by this interaction appears to play a critical role in the ability of these DNA viruses to transform cells, paralleling the similar action of MDM2. These interactions suggest that functional antagonism between oncogene and anti-oncogene products is likely to be critical for the control of cellular growth, with changes in this balance which activate oncogenes or inactivate anti-oncogenes resulting in cancer.

These considerations evidently focus attention on the genes that are activated by p53. One such gene is that encoding a 21-kDa protein (p21), which acts as an inhibitor of cyclin-dependent kinases (reviews: Peter and Herskowitz, 1994; Morgan, 1995). As the cyclin-dependent kinases are enzymes that stimulate cells to enter cell division, the finding that p53 stimulates the expression of an inhibitor of these enzymes is entirely consistent with its role in restraining growth, since the inhibition of the cyclin-dependent kinases will prevent cells replicating their DNA and undergoing cell division (review:Marx, 1993) (Figure 7.23). Interestingly, p53 also stimulates expression of the *mdm2* gene (Barak *et al.*, 1993), whose protein product interferes with the activity of p53, as described above. This effect is likely to be part of a negative feedback loop in which p53, having fulfilled its function, activates *mdm2* expression, resulting in p53 inactivation (Figure 7.24). This would allow, for example, cells that had repaired the damage to their DNA to inactivate p53 and resume cell division. Similarly, p53 also stimulates the expression of the bax gene, whose protein product stimulates programmed cell death or apoptosis, supporting a role for p53 in promoting the death of cells whose damaged DNA is irrepairable (Miyashita and Reed, 1995).

Hence the p53 gene product plays a key role in regulating cellular growth by binding to DNA and activating the expression of specific genes. Its inactivation by mutation or by interaction with oncogene products is likely to play a critical role in most human cancers.

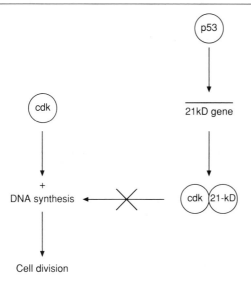

Figure 7.23 p53 activates the gene for the 21-kDa inhibitor of cyclic dependent kinases (cdk). This inhibitor then prevents the cyclin-dependent kinases from stimulating DNA synthesis and consequent cell division.

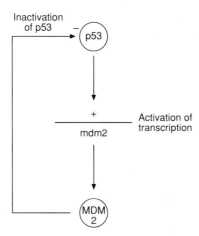

Figure 7.24 p53 activates the gene for the MDM2 protein, which acts in a negative feedback loop to inactivate p53 by the mechanism illustrated in Figure 7.22d.

7.3.3 The Retinoblastoma protein

The Retinoblastoma gene (Rb-1) was the first anti-oncogene to be defined and is so named because its inactivation in humans results in the formation of eye tumours known as retinoblastomas (reviews: Cobrinick *et al.*, 1992, Hamel *et al.*, 1992). Like p53, the Rb-1 gene product is a transcription factor which exerts its anti-oncogenic effect by modulating the expression of specific target genes. In contrast to p53, however, it exerts this effect via protein–protein interactions with other transcription factors rather than by direct DNA binding.

One of the major targets for Rb-1 is the transcription factor E2F, which plays a critical role in stimulating the expression of genes encoding growth-promoting proteins such as Myb (Section 7.2.3), Myc (Section 7.2.4), DNA polymerase α and thymidine kinase (review: Wagner and Green, 1991). The association of Rb-1 and E2F does not inhibit the DNA binding of E2F but prevents it from stimulating the transcription of these growth-promoting genes (Figure 7.25a). Hence Rb-1 exerts its anti-oncogenic effect by inhibiting the transcription of growth-promoting genes rather than, as with p53, promoting the transcription of growth-inhibiting genes.

In normal dividing cells, this interaction of Rb-1 and E2F is inhibited as cells move from G to S phase in the cell cycle. This effect is dependent on the phosphorylation of Rb-1, which prevents it interacting with E2F (Figure 7.25b). Hence the controlled growth of normal cells can be regulated by the regulated phosphorylation of Rb-1, which in turn regulates its ability to interact with E2F and modulate its activity. Interestingly, the phosphorylation of Rb-1 in the cell cycle is carried out by the cyclin-dependent kinases (review: Sherr, 1994). This provides a link between p53 and the regulation of Rb-1 activity since, as noted above (Section 7.3.2), p53 activates the gene encoding the p21 protein, which inhibits cyclin-dependent kinases, and would thus prevent the phosphorylation of Rb-1 and cell cycle progression.

Clearly, abolishing the activity of Rb-1, either by deletion of its gene or by mutation, will result in the unregulated activity of E2F, leading to the uncontrolled growth that is characteristic of cancer cells (Figure 7.25c). Interestingly, the inactivation of Rb-1 can also be achieved by the transforming proteins of DNA tumour viruses, such as SV40 or adenovirus. These proteins bind to the Rb-1 protein, resulting in the dissociation of the Rb-1–E2F complex, releasing free E2F which can activate gene expression (Figure 7.25d).

Although E2F is a major target of Rb-1, there are other factors with which Rb-1 interacts. Thus Rb-1 has recently been shown to inactivate the UBF factor, which plays a critical role in transcription of the ribosomal RNA genes by RNA polymerase I (see Section 3.2.2). Due to the need for these ribosomal RNAs for effective functioning of ribosomes, the inactivation of UBF by Rb-1 will lead to a decrease in the levels of total protein synthesis, which would in turn lead to the arrest of cell growth. In agreement with this idea, the inactivation of UBF by Rb-1 appears to play a critical role in the growth arrest and

(a) Resting cells

No transcription

(b) Dividing cells

Transcription

(c) Deletion or inactivation of Rb

Transcription

(d) Interaction with tumour virus oncogene

Transcription

Figure 7.25 In resting cells, the Rb-1 protein binds to E2F and prevents it activating the transcription of genes encoding growth stimulating proteins (GSGs) (a). In normal dividing cells, the Rb-1 protein is phosphorylated at the G1/S transition in the cell cycle, which prevents it from interacting with E2F and hence allows E2F to activate transcription (b). This release of E2F can also occur in tumour cells where the Rb-1 gene is deleted or inactivated by mutation (c) or following the interaction of Rb-1 with tumour virus oncogenes (T) (d).

associated differentiation of U937 monocytic cells (review: Dynlacht, 1995).

As well as these interactions with E2F and UBF, the unphosphorylated form of Rb-1 has been shown to interact with the muscle-determining factor MyoD (see Section 5.3) and to be necessary for the arrest of cell division that occurs during the terminal differentiation of muscle cells (Gu *et al.*, 1993). In addition, as discussed in Section 5.3.1, MyoD can activate the gene encoding the p21 inhibitor of cyclin dependent kinases (review: Marx, 1995), thereby preventing these kinases from phosphorylating Rb-1 (see above). Hence, as well as interacting with the dephosphorylated form of Rb-1, MyoD can also enhance the proportion of Rb-1 in the unphosphorylated form.

These findings have led to the suggestion that Rb-1 may act as a binary switch, regulating the balance between the continuation of cell division and the arrest of such division that is necessary for terminal differentiation. Thus the phosphorylated form of Rb-1 would promote cell division (see above) whilst its dephosphorylation would allow it to interact with specific transcription factors involved in terminal differentiation in specific cell types (such as MyoD in muscle cells), thereby promoting growth arrest and consequently differentiation (Figure 7.26). In agreement with this idea, mice in which the Rb-1 gene has been inactivated die before birth and show gross defects in cellular differentiation (Lee *et al.*, 1992). This indicates that Rb-1 plays a key role in normal development as well as acting as an anti-oncogene, and contrasts with the viability of mice in which the p53 gene has been inactivated (see Section 7.3.2).

Hence the Rb protein plays a key role in regulating cellular growth and differentiation by interacting with transcription factors involved in transcription by RNA polymerases I and II. Its inactivation either by mutation or by specific oncogenes therefore results in uncontrolled proliferation and cancer. When taken together with the similar role of p53 in growth regulation and as a target for oncogenes, this suggests that anti-oncogenes are likely to play a key role in regulating cellular growth, which is likely to be controlled by the balance between the antagonistic effects of oncogene and anti-oncogene products.

Figure 7.26 The Rb-1 protein may act as a switch mechanism, regulating the balance between cell growth and terminal differentiation. Thus, whilst phosphorylation of Rb-1 is necessary to allow cellular division, its dephosphorylation allows it to interact with factors such as MyoD and promote the cessation of cell division associated with terminal differentiation.

7.4 CONCLUSIONS

The ability to affect cellular transcriptional regulatory processes is crucial to the ability of many different viruses to transform cells. Thus, for example, the large T oncogenes of the small DNA tumour viruses SV40 and polyoma, and the Ela protein of adenovirus, can all affect cellular gene expression, and this ability is essential for the transforming ability of these viruses (review: Moran, 1993). In this chapter we have seen that several RNA viruses also have this ability, containing transcription factors that can act as oncogenes either by promoting the expression of genes required for growth or for the maintenance of the proliferating undifferentiated state or by inhibiting the expression of genes required for the production of non-proliferating differentiated cells.

Although the oncogenes of both DNA and some RNA tumour viruses can therefore affect transcription, their origins are completely different. Thus, whilst the oncogenes of the DNA viruses do not have equivalents in cellular DNA and appear to have evolved within the viral genome, the oncogenes of retroviruses have, as we have seen, been picked up from the cellular genome. The fact that, despite their diverse origins, both types of oncogenes can affect transcription, indicates that the modulation of transcription represents an effective mechanism for the transformation of cells.

In addition, however, the origin of retroviral oncogenes from the cellular genome allows several other features of transcription to be studied. Thus, for example, the conversion of a normal cellular transcription factor into a cancer-causing viral oncogene allows insights to be obtained into the processes whereby oncogenes become activated.

In general, such oncogenes, whether they encode growth factors, growth factor receptors or other types of protein, can be activated within a virus either by over-expression driven by a strong retroviral promoter or by mutation. The transcription factors we have discussed in this chapter illustrate both these processes. Thus the *fos, jun* and *myc* oncogenes, for example, become cancer-causing both by continuous expression of proteins that are normally made only transiently, leading to constitutive stimulation of genes required for growth, as well as in some cases by mutations in the viral forms of the protein which render them more potent transcriptional activators. Similarly, the *erbA* and *myb* oncogenes are activated by deletion of a part of the protein-coding regions, leading to a protein with different or enhanced properties. Although such effects of mutation or over-expression have initially been defined in tumorigenic retroviruses, it is likely such changes can also occur within the cellular genome, over-expression of the c-*myc* oncogene, for example, being characteristic of many different human tumours (review: Spencer and Groudine, 1991), whilst several other transcription factor genes are activated by the translocations characteristic of particular human leukaemias (see Section 7.2.5).

In addition, since cellular oncogenes clearly also play an important role in the regulation of normal cellular growth and differentiation, their identification via tumorigenic retroviruses has, paradoxically, greatly aided the study of normal cellular growth regulatory processes. Thus, for example, the prior isolation of the c-*fos* and c-*jun* genes greatly aided the characterization of AP1-binding activity and of its role in stimulating genes involved in cellular growth. Similarly, the c-*myb* gene, whose product appears to play a critical role in gene regulation during monocyte differentiation, was initially identified by the isolation of the v-myb gene from a retrovirus.

A similar boost to our understanding of growth regulation in normal cells has also emerged from studies of the anti-oncogene proteins. Thus studies on the Rb-1 gene, which was orginally identified on the basis of its inactivation in retinoblastomas, have led to an understanding of its key role in regulating the balance between cellular growth and differentiation. Similarly, work on p53, which was originally identified as a protein interacting with the product of the SV40 large T oncogene, has led to the identification of its key role as the so-called 'guardian of the genome'.

The interaction of p53 and SV40 large T indicates another aspect of anti-oncogenes, namely their antagonistic interaction with oncogene products. Thus both p53 and Rb-1 have been shown to bind cellular and viral oncogene proteins, with the activity of the anti-oncogene product being inhibited by this interaction. Such interactions are not confined to oncogenes and anti-oncogenes that encode transcription factors. Thus the NF-1 anti-oncogene encodes a protein that activates the GTP-hydrolysing ability of the *ras* oncogene product, so switching it to the non-growth-promoting form. Hence the interaction between oncogene and anti-oncogene products is likely to play a key role in regulating cellular growth. The uncontrolled growth characteristic of cancer cells therefore results from changes in this balance, due either to over-expression or mutational activation of oncogenes or to deletion or mutational inactivation of anti-oncogenes.

It is clear, therefore, that, as with other oncogenes and anti-oncogenes, the study of the oncogenes and anti-oncogenes that encode transcription factors can provide considerable information on both the processes regulating normal growth and differentiation and on how these processes are altered in cancer.

REFERENCES

Amati, B., Brooks, M. W., Levy, N., Littlewood, T. D., Evan, G. I. and Land, H. (1993). Oncogenic activity of the c-Myc protein requires dimerization with Max. *Cell* **72**, 233–245.

Amati, B. and Land, H. (1994). Myc-Max-Mad: a transcription factor network controlling differentiation and death. *Current Opinion in Genetics and Development* **4**, 102–108.

Anderson, M. E. and Tegtmeyer, P. (1995). Giant leap for p53, small step for drug design. *Bioessays* **17**, 3–7.

Angel, P., Allegretto, E. A., Okino, S. T., Hattori, K., Boyle, W. J., Hunter, T. and Karin, M. (1988). Oncogene *jun* encodes a sequence-specific trans-activator similar to AP-1. *Nature* **332**, 166–170.

Baichwal, U. R. and Tjian, R. (1990). Control of c-Jun activity by interaction of a cell-specific inhibitor with regulatory domain delta: differences between v- and c-Jun. *Cell* **63**, 815–825.

Barak, Y., Juven, T., Haffner, R. and Oren, M. (1993). mdm2 expression is induced by wild type p53 activity. *EMBO Journal* **12**, 461–468.

Berns, A. (1994). Is p53 the only real tumour suppressor gene? *Current Biology* **4**, 137–139.

Biedenkapp, H., Borgmeyer, U., Sippel, A. E. and Klempnauer, K-H. (1988). Viral *myb* encodes a sequence-specific DNA-binding activity. *Nature* **335**, 835–837.

Bishop, J. M. (1987). The molecular genetics of cancer. *Science* **235**, 305–311.

Blackwood, E. M. and Eisenman, R. N. (1991). Max: a helix–loop–helix zipper protein that forms a sequence-specific DNA binding complex with myc. *Science* **251**, 1211–1217.

Blackwood, E. M., Kretzner, L. and Eisenman, R. N. (1992). Myc and max function as a nucleoprotein complex. *Current Opinion in Genetics and Development* **2**, 227–235.

Bourne, H. R. and Varmus, H. E. (eds) (1992). Oncogenes and cell proliferation. *Current Opinion in Genetics and Development* **2**, 1–57.

Bours, V., Franzoso, G., Azarenko, V., Parks Karno, T., Brown, K. and Siebenlist, U. (1993). The oncoprotein Bcl-3 directly transactivates through κB motifs via association with DNA-binding p50B homodimers. *Cell* **72**, 729–739.

Cobrinick, D., Dowdy, S. F., Hinds, P. W., Mittnacht, S. and Weinberg, R. A. (1992). The retinoblastoma protein and the regulation of cell cycling. *Trends in Biochemical Sciences* **17**, 312–315.

Curran, T. and Franza, B. R. (1988). Fos and Jun: the AP-1 connection. *Cell* **55**, 315–397.

Downward, J., Yarden, Y., Mayes, E., Scrace, G., Totty, N., Stockwell, P.,

Ullrich, A., Schlessinger, J. and Watefield, M. D. (1984). Close similarity of epidermal growth factor receptor and v-*erb-B* oncogene protein sequences. *Nature* **307**, 521–527.

Dyck, J. A., Maul, G. G., Miller, W. H., Chen, J. D., Kakizuka, A. and Evans, R. M. (1994). A novel macromolecular structure is a target of the promyelocyte–retinoic acid receptor oncoprotein. *Cell* **76**, 333–343.

Dynlacht, B. D. (1995). PolI gets repressed. *Nature* **374**, 114.

Forrest, D. and Curran, T. (1992). Crossed signals: oncogenic transcription factors. *Current Opinion in Genetics and Development* **2**, 19–27.

Friend, S. (1994) p53: a glimpse at the puppet behind the shadow play. *Science* **265**, 334–335.

Gilmore, T. D. (1990). NF-kappaB, KBF1, *dorsal* and *rel*ated matters. *Cell* **62**, 841–843.

Graf, T. (1992). Myb, a transcriptional activator linking proliferation and differentiation in haematopoetic cells. *Current Opinion in Genetics and Development* **2**, 249–255.

Gu, W., Schneider, J. W., Condorelli, G., Kaushall, S., Mahdavi, V. and Nadal-Ginard, B. (1993). Interaction of myogenic factors and the retinoblastoma protein mediates muscle cell commitment and differentiation. *Cell* **72**, 309–324.

Gutman, A. and Wasylyk, B. (1991). Nuclear targets for transcription regulation by oncogenes. *Trends in Genetics* **7**, 49–54.

Halazonetis, T. D., Georgopoulos, K., Greenberg, M. E. and Leder, P. (1988). C-*jun* dimerizes with itself and with c-*fos* forming complexes of different DNA binding abilities. *Cell* **55**, 917–924.

Hamel, P. A., Gallie, B. L. and Phillips, R. A. (1992). The retinoblastoma protein and cell cycle regulation. *Trends in Genetics* **8**, 180–185.

Hastie, N. D. (1993). Wilms' tumour gene structure and function. *Current Opinion in Genetics and Development* **3**, 408–413.

Johnson, R. S., Van Lingen, B., Papaivannou, V. E. and Speigelmann, B. M. (1993). A null mutation at the c-*jun* locus causes embryonic lethality and retarded cell growth in culture. *Genes and Development* **7**, 1309–1317.

Kerppola, T. and Curran, T. (1995). Zen and the art of Fos and Jun. *Nature* **373**, 199–200.

Knudson, A. G. (1993). Anti-oncogenes and human cancer. *Proceedings of the National Academy of Sciences USA* **90**, 10914–10921.

Kretzner, L., Blackwood, E. M. and Eisenmann, R. N. (1992). Myc and Max proteins possess distinct transcriptional activities. *Nature* **359**, 426–428.

Lamph, W. W., Wamsley, P., Sassone-Corsi, P. and Verma, I. M. (1988). Induction of proto-oncogene Jun/AP–1 by serum and TPA. *Nature* **334**, 629–631.

Land, H., Parada, L. F. and Weinberg, R. (1983). Cellular oncogenes and multistep carcinogenesis. *Science* **222**, 771–778.

Lane, D. P. (1992). p53, guardian of the genome. *Nature* **358**, 15–16.

Lane, T., Ibanez, C., Garcia, A., Graf, T. and Lipsick, J. C. (1990). Transformation by v-*myb* correlates with trans-activation of gene expression. *Molecular and Cellular Biology* **10**, 2491–2598.

Lee, E. Y-H. P., Chang, C. Y., Hu, N., Wang, Y. C., Lai, C. C., Herrup, K., Lee, W. H. and Bradley, A. (1992). Mice deficient for Rb are non viable and show defects in neurogenesis and haematopoesis. *Nature*, **359**, 289–294.

Levine, A. J., Momaid, J. and Finlay, C. A. (1991). The p53 tumour suppressor gene. *Nature* **351**, 453–456.

Li, L., Chambard, J. C., Karin, M. and Olson, E. M. (1992). Fos and Jun repress transcriptional activation by myogenin and MyoD: the amino terminus of Jun can mediate repression. *Genes and Development* **6**, 676–689.

Luscher, B., Christenson, E., Litchfield, D. W., Krebs, E. G. and Eisenman, R. N. (1990). *Myb* DNA binding inhibited by phosphorylation at a site deleted during oncogenic activation. *Nature* **344**, 517–522.

Luscher, B. and Eisenman, R. N. (1990). New light on Myc and Myb. Part II, Myb. *Genes and Development* **4**, 2235–2241.

Marx, J. (1993). How p53 suppresses cell growth. *Science* **262**, 1644–1645.

Marx, J. (1995). Cell cycle inhibitors may help brake growth as cells develop. *Science* **267**, 963–964.

Miyashita, T. and Reed, J. C. (1995). Tumor suppressor p53 is a direct transcriptional activator of the human *bax* gene. *Cell* **80**, 293–299.

Moran, E. (1993). DNA tumour virus transforming proteins and the cell cycle. *Current Opinion in Genetics and Development* **3**, 63–70.

Morgan, D. O. (1995). Principles of CDK regulation. *Nature* **374**, 131–134.

Ness, S. A., Beug, H. and Graf, T. (1987). v-*myb* dominance over v-*myc* in doubly transformed myelomonocytic cells. *Cell* **51**, 41–50.

Ness, S. A., Kowenz-Lentz, E., Casini, T., Graf, T. and Leutz, A. (1993). Myb and NF-M: combinatorial activators of myeloid genes in heterologous cell types. *Genes and Development* **7**, 749–759.

Ness, S. A., Marknell, A. and Graf, T. (1989). The v-*myb* oncogene product binds to and activates the promyelocyte-specific *mim*-1 gene. *Cell* **59**, 1115–1125.

Oliner, J. D., Pieterpol, J. A., Thiagalingam, S., Gyuris, J., Kinzler, K. W. and Vogelstein, B. (1993). Oncoprotein MDM2 conceals the activation domain of tumour suppressor p53. *Nature* **362**, 857–860.

Peter, M. and Herskowitz, I. (1994). Joining the complex: cyclin dependent kinase inhibitory proteins and the cell cycle. *Cell* **79**, 181–184.

Rabbits, T. H. (1994). Chromosomal translocations in human cancer. *Nature* **372**, 143–149.

Ransone, L. J. and Verma, I. M. (1990). Nuclear proto-oncogenes. Fos and Jun. *Annual Review of Cell Biology* **6**, 531–557.

Rauscher, F. J., Cohen, D. R., Curran, T., Bos, T. J., Bogt, P. K., Bohmann, D., Tjian, R. and Franza, B. R. (1988). Fos-associated protein p39 is the product of the c-*jun* oncogene. *Science* **240**, 1010–1016.

Reith, W., Satola, S., Sanchey, C. H., Amaldi, I., Lisowska-Grospiere, B., Griscelli, C., Hadam, M. R. and Mach, B. (1988). Congenital immunodeficiency with a regulatory defect in MHC class II gene expression lacks a specific HLA-DR promoter binding protein RF-X. *Cell* **53**, 897–906.

Sap, J., Munoz, A., Damm, K., Goldberg, Y., Ghysdael, J., Leutz, A., Beug, H. and Vennstrom, B. (1986). The c-*erb-A* protein is a high-affinity receptor for thyroid hormone. *Nature* **324**, 635–640.

Sap, J., Munoz, A., Schmitt, A., Stunnenberg, H. and Vennstrom, B. (1989). Repression of transcription at a thyroid hormone response element by the v-*erbA* oncogene product. *Nature* **340**, 242–244.

Sawyers, C. L. and Denny, C. T. (1994). Chronic myelomonocytic leukaemia. Tel-a-kinase what ets all about. *Cell* **77**, 171–173.

Schuermann, M., Neuberg, M., Hunter, J. B., Henuwein, T., Ryseck, R-P., Bravo, R. and Muller, R. (1989). The leucine repeat motif in *fos* protein mediates complex formation with Jun/AP1 and is required for transformation. *Cell* **56**, 507–516.

Sherr, C. J. (1994). G1 phase progression: cycling on cue. *Cell* **79**, 551–555.

Spencer, C. A. and Groudine, M. (1991). Control of c-*myc* regulation in normal and neoplastic cells. *Advances in Cancer Research* **56**, 1–48.

Treier, M., Staszewski, L. M. and Bohmann, D. (1994). Ubiquitin-dependent c-Jun degradation *in vivo* is mediated by the δ domain. *Cell* **78**, 787–798.

Vogelstein, B. and Kinzler, K. W. (1992). p53 function and dysfunction. *Cell* **70**, 523–526.

Vogt, P. K., Bos, T. J. and Doolittle, R. F. (1987). Homology between the DNA-binding domain of the GCN4 regulatory protein of yeast and the carboxyl terminal region of a protein coded for by the oncogene *jun*. *Proceedings of the National Academy of Sciences USA* **84**, 3316–3319.

Wagner, S. and Green, M. R. (1991). A transcriptional tryst. *Nature* **352**, 189–190.

Wasylyk, B., Wasylyk, C., Flores, P., Begue, A., Leprince, D. and Stehelin, D. (1990). The c-*ets* proto-oncogenes encode transcription factors that copoperate with c-Fos and c-Jun for transcriptional activation. *Nature* **346**, 191–193.

Weinberg, R. A. (1993). Tumour supressor genes. *Neuron* **11**, 191–196.

Weinberger, C., Thompson, C. C., Ong, E. S., Lebo, R., Gruol, D. J. and Evans, R. M. (1986). The c-*erb*-A gene encodes a thyroid hormone receptor. *Nature* **324**, 641–646.

Weston, I. X. and Bishop, J. M. (1989). Transcriptional activation by the v-*myb* oncogene and its cellular progenitor c-*myb*. *Cell* **58**, 85–93.

Zenke, M., Kahn, D., Disela, C., Vennstrom, B., Leutz, A., Keegan, K., Hayman, M. J., Choe, H-R., Yew, N., Engel, J. D. and Beug, H. (1988). v-*erb* A specifically suppresses transcription of the avian erythrocyte anion transporter (Band 3) gene. *Cell* **52**, 107–119.

CHAPTER EIGHT

DNA binding by transcription factors

8.1 INTRODUCTION

In previous chapters we have considered the role of transcription factors in various processes such as constitutive, inducible and tissue-specific gene expression. To fulfil these roles, transcription factors must possess certain features allowing them to modulate gene expression. These features will be discussed in the remaining chapters of this book. Clearly, the first feature that these factors require is the ability to bind to DNA in a sequence-specific manner and this is discussed in this chapter. Following binding, the factor must interact with other factors or with the RNA polymerase itself in order to influence transcription either positively or negatively, and this aspect is discussed in Chapter 9. Finally, in the case of factors modulating inducible, tissue-specific or developmentally regulated gene expression, some means must exist to regulate the synthesis or activity of the factor so that it is active only in a particular situation. This regulation of factor synthesis or activity is discussed in Chapter 10.

Following the cloning of many different eukaryotic transcription factors, the domain-mapping experiments described in Section 2.3.3 have led to the identification of several distinct structural elements in different factors which can mediate DNA binding. These motifs will be discussed in turn, using transcription factors that contain them to illustrate their properties (reviews: Harrison, 1991; Pabo and Sauer, 1992; Travers, 1993).

8.2 THE HELIX–TURN–HELIX MOTIF IN THE HOMEOBOX

As discussed in Section 6.2.2, the DNA-binding abilities of the homeo-box-containing transcription factors are mediated by the homeobox region of the protein. Thus this region, when synthesized without the remainder of the protein, either by expression in bacteria or by chemical synthesis, can bind to DNA in the identical sequence-specific manner exhibited by the intact protein.

This ability to define the 60-amino-acid homeodomain as the region binding to DNA has led to intensive study of its structure in the hope of elucidating how the protein binds to DNA in a sequence-specific manner (reviews: Kornberg, 1993; Gehring *et al.*, 1994). In particular, the crystal structure of the Antennapedia (Antp) homeodomain bound to DNA has been determined by nuclear magnetic resonance (NMR) spectroscopy, whilst similar structural studies of the engrailed (Eng) and MATα2 homeodomains bound to DNA have been carried out by X-ray crystallography.

By this means, the Antp homeodomain was shown to contain a short N-terminal arm of six residues followed by four α-helical regions (Figure 8.1). The first two helices are virtually anti-parallel to each other, with the other two helices arranged at right angles to the first. Most inter-estingly, helices II and III are separated by a β-turn, forming a helix–turn–helix motif (Figure 8.2). The Eng and MATα2 homeodomains also have a similar structure, with an N-terminal arm and a subsequent

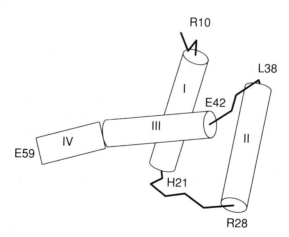

Figure 8.1 Structure of the Antennapedia homeodomain as determined by nuclear magnetic resonance spectroscopy. Note the four α-helical regions (I–IV) represented as cylinders, with the amino acids at their ends indicated by numbers and the one-letter amino acid code.

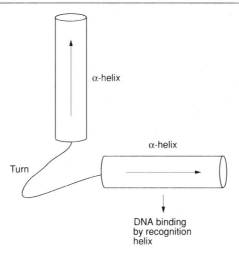

Figure 8.2 The helix–turn–helix motif.

helix–turn–helix motif. In this case, however, the third and fourth helices observed in Antp form a single helical region. Interestingly, the helix–turn–helix stucture typical of the homeodomain is very similar to the DNA-binding motif of several bacteriophage regulatory proteins, such as the lambda cro protein or the phage 434 repressor, which have also been crystallized and subjected to intensive structural study.

In these bacteriophage proteins X-ray crystallographic studies have shown that the helix–turn–helix motif does indeed contact DNA. One of the two helices lies across the major groove of the DNA, whilst the other lies partly within the major groove, where it can make sequence-specific contacts with the bases of DNA. It is this second helix (known as the recognition helix) that therefore controls the sequence-specific DNA-binding activity of these proteins (Figure 8.3).

The similarity in structure of helices II and III in the eukaryotic homeodomains with the two helices of the bacteriophage proteins led to the suggestion that these two helices in the homeodomain are similarly aligned relative to the DNA, with helix III constituting the recognition helix responsible for sequence-specific DNA binding. Hence the precise amino acid sequence in the recognition helix in different homeodomain proteins would determine the DNA sequence that they bound (review: Treisman *et al.*, 1992).

In agreement with this idea, exchanging the recognition helix in the Bicoid (Bcd) homeodomain for that of Antp resulted in a protein with the DNA-binding specificity of Antp and not of Bicoid. Most interestingly, a Bcd protein with the DNA-binding specificity of Antp could also be obtained by exchanging only the ninth amino acid in the recognition

Figure 8.3 Binding of the helix–turn–helix motif to DNA with the recognition helix in the major groove of the DNA.

helix, replacing the lysine residue in Bcd with the glutamine residue found in the Antp protein (Figure 8.4), whereas the exchange of other residues that differ between the two proteins has no effect on the DNA-binding specificity. Hence the ninth amino acid within the recognition helix of the homeodomain plays a critical role in determining DNA-binding specificity.

It is likely that the amino group of lysine found at the ninth position in the Bcd protein makes hydrogen bonds with the O6 and N7 positions of a guanine residue in the Bcd-specific DNA-binding site, whereas the amide group of glutamine found at the corresponding position in the

	Recognition helix										Binding to	
											Bicoid site TCTAATCCC	Antp site TCAATTAAAT
Bicoid	T	A	Q	V	K	I	W	F	K	N	+	−
	A	−	−	−	−	−	−	−	−	−	+	−
	−	−	−	−	A	−	−	−	−	−	+	−
	−	−	−	−	−	−	−	−	A	−	−	−
	A	−	−	−	A	A	−	−	A	−	−	−
	E	R	−	−	−	−	−	−	Q	−	−	+
	E	R	−	−	−	−	−	−	−	−	+	−
	−	−	−	−	−	−	−	−	Q	−	−	+
Antp	E	R	Q	I	K	I	W	F	Q	N	−	+

Figure 8.4 Effect of changing the amino acid sequence in the recognition helix of the Bicoid protein on its binding to its normal recognition site and that of the Antennapedia (Antp) protein. Note the critical effect of changing the ninth amino acid in the helix, which completely changes the specificity of the Bicoid protein.

Antp recognition helix forms hydrogen bonds with the N6 and N7 positions of an adenine residue at the equivalent position within the Antp-specific DNA-binding site. Hence the replacement of lysine with glutamine results in the loss of two potential hydrogen bonds to a Bcd site and the gain of two potential hydrogen bonds to an Antp site, explaining the observed change in DNA-binding specificity (Figure 8.5).

A similar critical role for the ninth amino acid in determining the precise DNA sequence that is recognized is also seen in other homeobox-containing proteins, replacement of the serine found at this position in the paired protein with the lysine found in Bicoid or the glutamine found in Antp allowing the paired protein to recognize respectively Bcd- or Antp-specific DNA sequences. Hence the DNA sequence recognized by

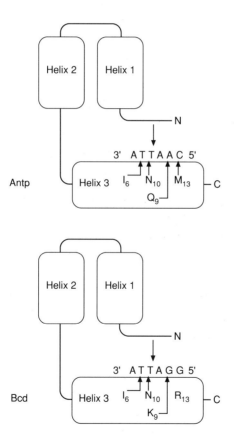

Figure 8.5. Contacts between DNA and the Antp or Bcd homeodomains. Note that the changes in the ninth amino acid of the recognition helix (helix 3) alters the base that is preferentially bound from G for Bcd to A for Antp, as discussed in the text, whilst the N-terminal arm of the homeodomain contacts the ATTA sequence common to the recognition site of both proteins.

a homeobox-containing protein appears to be determined primarily by the ninth amino acid in the recognition helix, proteins with different amino acids at this position recognizing different DNA sequences, whereas proteins such as Antp and Fushi-tarazu, which have the same amino acid at this position, recognize the same DNA sequence.

This critical role of the ninth amino acid is in contrast to the situation in the bacteriophage proteins, in which the helix–turn–helix motif was originally defined. In these proteins, the most N-terminal residues (1–3) in the recognition helix play a critical role in determining DNA-binding specificity (review: Pabo and Sauer, 1992). As shown in Figure 8.4, however, these amino acids appear to play little or no role in determining the DNA-binding specificity of eukaryotic helix–turn–helix proteins, suggesting, therefore, that the recognition helix of these proteins is oriented differently in the major groove of the DNA.

This idea is in agreement with structural studies of the eukaryotic homeodomains bound to DNA, which have identified the actual protein–DNA contacts. These studies have shown that, as in the bacteriophage proteins, the recognition helix directly contacts the bases of DNA in the major groove. However, in the eukaryotic homeobox proteins, this helix is oriented within the major groove somewhat differently, such that the critical base-specific contacts are, as predicted, made by the C-terminal end of the helix, which contains residue nine (Figure 8.5).

It is clear, therefore, that the helix–turn–helix motif in the homeobox mediates both the DNA binding of the protein and also, via the recognition helix, controls the precise DNA sequence that is recognized. Interestingly, however, the short N-terminal arm of the homeodomain also contacts the bases of the DNA, although it makes contact in the minor groove rather than the major groove. Removal of this short N-terminal arm dramatically reduces the DNA-binding affinity of the homeodomain, indicating that this region contributes significantly to the DNA-binding ability of the homeodomain, probably by contacting the ATTA bases common to the DNA-binding sites of several homeodomain proteins (Figure 8.5).

This critical role for the helix–turn–helix motif in the homeodomain is also seen in the POU proteins which, as discussed in Section 6.3.2, contain the homeobox as part of a much larger conserved domain, which also contains a POU-specific region. The recent determination of the crystal structure of the Oct-1 POU domain bound to DNA (Klemm *et al.*, 1994) has shown that the Oct-1 homeodomain binds in a similar manner to the classical homeobox proteins, with the recognition helix lying in the major groove and the N-terminal arm in the minor groove. As noted in Chapter 6, however, whilst the isolated homeodomain can bind to DNA in a sequence-specific manner, both the binding affinity and sequence specificity are greatly increased in the presence of

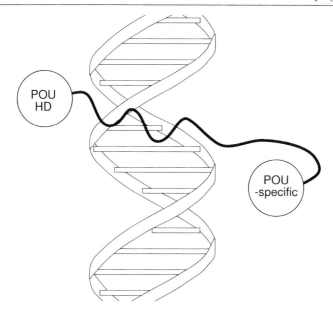

Figure 8.6. Binding of the POU-specific domain and POU homeodomain to opposite sides of the DNA double helix. Note the flexible linker region joining the two DNA-binding motifs.

the POU-specific domain, which is separated from the homeodomain by a short flexible linker sequence. Like the homeodomain, the POU-specific domain forms a helix–turn–helix motif, which allows it to bind to bases within the DNA adjacent to those contacted by the homeodomain, with binding of the two regions occurring on opposite sides of the DNA double helix (Figure 8.6).

8.3 THE ZINC FINGER MOTIF

8.3.1 The two-cysteine two-histidine finger

Transcription factor TFIIIA plays a critical role in regulating the transcription of the 5S ribosomal RNA genes by RNA polymerase III (see Section 3.2.3). When this transcription factor was purified, it was found to have a repeated structure and to be associated with between 7 and 11 atoms of zinc per molecule of purified protein (Miller *et al.*, 1985). When the gene encoding TFIIIA was cloned, it was shown that this repeated structure consisted of the unit, Tyr/Phe-X-Cys-X-Cys-$X_{2,4}$-Cys-X_3-Phe-

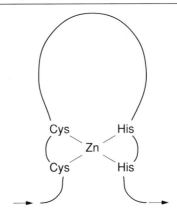

Figure 8.7 Schematic representation of the zinc finger motif. The finger is anchored at its base by the conserved cysteine and histidine residues, which tetrahedrally coordinate an atom of zinc.

X_5-Leu-X_2-His-$X_{3,4}$-His-X_5, which is repeated nine times within the TFIIIA molecule. This repeated structure therefore contains two invariant cysteine and two invariant histidine residues, which were predicted to bind a single zinc atom, accounting for the multiple zinc atoms bound by the intact molecule.

This motif is referred to as a zinc finger on the basis of its proposed structure in which a loop of 12 amino acids containing the conserved leucine and phenylalanine residues as well as several basic amino acids projects from the surface of the molecule, being anchored at its base by the cysteine and histidine residues that tetrahedrally co-ordinate an atom of zinc (Figure 8.7). The proposed interaction of zinc with the conserved cysteine and histidine residues in this structure was subsequently confirmed by X-ray adsorption spectroscopy of the purified TFIIIA protein (Diakun *et al.*, 1986).

Following its identification in the RNA polymerase III transcription factor TFIIIA, similar Cys$_2$-His$_2$-containing zinc finger motifs were identified in a number of RNA polymerase II transcription factors, such as Sp1, which contains three contiguous zinc fingers (Kadonaga *et al.*, 1987) and the *Drosophila* Kruppel protein, which contains four finger motifs (Rosenberg *et al.*, 1986; see Section 6.1). A list of zinc finger-containing transcription factors is given in Table 8.1 (reviews: Evans and Hollenberg, 1988; Struhl, 1989).

In all cases studied, the zinc finger motifs have been shown to constitute the DNA-binding domain of the protein, with DNA binding being dependent upon their activity. Thus, in the case of TFIIIA, DNA binding is dependent on the presence of zinc, allowing the finger structures to form (Hanas *et al.*, 1983), whilst progressive deletion of more and more

Table 8.1 Transcriptional regulatory proteins containing
Cys$_2$—His$_2$ zinc fingers

Organism	Gene	Number of fingers
Drosophila	Kruppel	4
	Hunchback	6
	Snail	4
	Glass	5
Yeast	ADR1	2
	SW15	3
Xenopus	TFIIIA	9
	Xfin	37
Rat	NGF-1A	3
Mouse	MK1	7
	MK2	9
	Egr1	3
	Evi1	10
Human	Sp1	3
	TDF	13

zinc finger repeats in the molecule results in a parallel loss of DNA-binding activity (Vrana *et al.*, 1988). Similarly, in the case of Sp1, DNA binding is dependent on the presence of zinc, and most importantly the sequence-specific binding activity of the intact protein can be reproduced by a protein fragment containing only the zinc finger region (Kadonaga *et al.*, 1987).

A similar dependence of DNA binding on the zinc finger motif is also seen in the *Drosophila* Kruppel protein which, as discussed in Chapter 6, is essential for correct thoracic and abdominal development. In this case a single mutation in one of the conserved cysteine residues in the finger, replacing it with a serine which cannot bind zinc, results in the production of a mutant fly indistinguishable from that produced by a complete deletion of the gene (Redemann *et al.*, 1988), indicating the vital importance of the zinc finger (Figure 8.8).

As with the helix–turn–helix motif of the homeobox, therefore, the zinc finger motif forms the DNA-binding element of the transcription factors that contain it. Interestingly, however, a single zinc finger taken from the yeast ADRI protein is unable to mediate sequence-specific DNA binding in isolation, whereas a protein fragment containing both the two fingers present in the intact protein can do so (Parraga *et al.*, 1988). This suggests, therefore, that DNA binding by the zinc finger is dependent upon interactions with adjacent fingers and explains why zinc finger-containing transcription factors always contain multiple copies of the zinc finger motif (see Table 8.1).

The initial structure proposed for the zinc finger, in which the zinc-coordinating cysteines and histidines form a base from which a loop of basic amino acids projects, provides an obvious basis for the DNA-

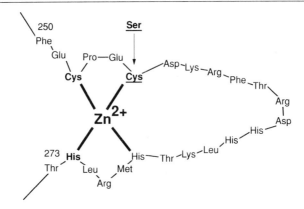

Figure 8.8 Structure of a zinc finger in the *Drosophila* Kruppel protein indicating the cysteine to serine change which abolishes the ability to bind zinc and results in a mutant fly indistinguishable from that obtained when the entire gene is deleted.

binding ability of the zinc finger. Thus, in the case of TFIIIA, it was proposed that the tips of the fingers, which were rich in basic amino acids such as arginine and lysine, would directly contact the acidic DNA in the major groove, with successive fingers binding on opposite sides of the helix (Figure 8.9) (reviews: Klug and Rhodes, 1987; Rhodes and Klug, 1993). This structure, in which each finger makes contact with adjacent regions of DNA, would allow the relatively small TFIIIA molecule to make contact, via its multiple fingers, with the relatively large 50-base-pair regulatory region of the 5*S* molecule.

Such a structure for the zinc finger, in which the zinc coordination via cysteine and histidine serves as a scaffold for the intervening region which makes direct contact with the DNA, is now generally accepted. However, it is unlikely that the intervening amino acids form a simple loop structure, as proposed in the original model. Thus an alternative view of the structure of the zinc finger has been inferred from the structures of other metalloproteins (Figure 8.10; Berg, 1988) and is

Figure 8.9 Model of the binding of the zinc fingers in TFIIIA to the 5*S* DNA. Note that adjacent fingers make contact with the DNA from opposite sides of the helix.

Figure 8.10 Alternative structure for the zinc finger in which two anti-parallel β-sheets (straight lines) are packed against an adjacent α-helix (wavy line).

supported by structural studies on the zinc fingers in the yeast SWI5 transcription factor (see Section 5.4.2) using NMR spectroscopy (Neuhaus *et al.*, 1990). In this structure, the finger region forms a motif consisting of two anti-parallel β-sheets with an adjacent α-helix packed against one face of the β-sheet. Upon contact with DNA, the α-helix would lie in the major groove of the DNA and make sequence-specific contacts with the bases of DNA, whilst the β-sheets lie further away from the helical axis of the DNA and contact the DNA backbone.

Most interestingly, this proposed structure suggests that a critical role in sequence-specific DNA binding will be played by amino acids at the N-terminus of the α-helix, most notably the amino acids immediately preceeding the first histidine residue. In agreement with this idea, two amino acids in this region play a critical role in determining the DNA-binding specificity of the *Drosophila* Krox-20 transcription factor (Nardelli *et al.*, 1991). Thus this factor contains three zinc fingers and interacts with the DNA sequence 5'-GCGGGGGCG-3'. If each finger contacts three bases within this sequence, then the central finger must recognize the sequence GGG, whereas the two outer fingers will each recognize the sequence GCG (Figure 8.11). When the amino acid sequence of each of the Krox-20 fingers was compared, it was found that the two outer fingers contain a glutamine residue at position 18 of the finger and an arginine at position 21, whereas the central finger has histidine and threonine residues at these positions. As expected, if these two amino acid differences are critical in determining the DNA sequence that is recognized, altering these two residues in the central finger to their equivalents in the outer two fingers resulted in a factor that failed

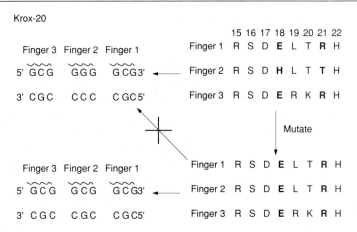

Figure 8.11 DNA-binding specificity and amino acid sequence of the three cysteine–histidine zinc fingers in the *Drosophila* Krox 20 protein. Note that each finger binds to three specific bases in the recognition sequence and that finger 2, which differs from fingers 1 and 3 in the DNA sequence it recognizes, also differs in the amino acids at positions 18 and 21 in the finger (bold letters). Mutating these amino acids to their equivalents in fingers 1 and 3 changes the DNA-binding specificity of finger 2 to that of fingers 1 and 3, indicating that these amino acids play a critical role in determining the DNA sequence that is recognized.

to bind to the normal Krox-20-binding site but instead bound to the sequence 5′-GCGGCGGCG-3′, in which each finger binds the sequence GCG.

This experiment therefore indicates the critical role of two amino acids at the N-terminus of the α-helix in producing the DNA-binding specificity of zinc fingers of this type and also shows that, at least in the case of Krox-20, each successive finger interacts with three bases of DNA within the recognition sequence. The importance of these amino acids has also been confirmed in recent experiments in which the amino acids at different positions in the zinc finger were randomly altered and their interaction with a wide range of DNA sequences assessed (Choo and Klug, 1994; Rebar and Pabo, 1994). Clearly, such an important role for the amino acids at the N-terminus of an α-helix parallels the similar critical role for the equivalent amino acids in the recognition helix of the bacteriophage DNA recognition proteins (see Section 8.2).

Hence, like the helix–turn–helix motif, the cysteine–histidine zinc finger plays a critical role in mediating the DNA-binding abilities of transcription factors that contain it, with sequence-specific recognition of DNA being determined in both cases by amino acids within an α-helix.

8.3.2 The multi-cysteine zinc finger

A similar zinc-binding domain to that discussed above has also been identified in the DNA-binding domains of the members of the steroid–thyroid hormone receptor family (see Section 4.3 and Parker (1993) for a discussion of these receptors). As with the cysteine–histidine fingers, this motif has been shown by X-ray adsorption spectroscopy to bind zinc in a tetrahedral configuration (Freedman *et al.*, 1988). However, in this case, coordination is achieved by four cysteine residues rather than the two-cysteine two-histidine structure discussed above. Similar multi-cysteine motifs have also been identified in several other DNA-binding transcription factors, such as the yeast proteins GAL4, PRRI and LAC9, as well as in the adenovirus transcription factor E1A (Table 8.2) (review: Evans and Hollenberg, 1988), indicating that this type of motif is not confined to the steroid–thyroid hormone receptors.

In the case of the steroid–thyroid hormone receptors, the DNA-binding domain has the consensus sequence $Cys-X_2-Cys-X_{13}-Cys-X_2-Cys-X_{15,17}-Cys-X_5-Cys-X_9-Cys-X_2-Cys-X_4-Cys$. This motif is therefore

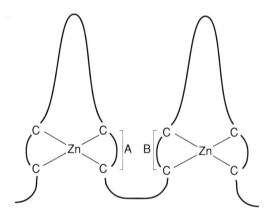

Figure 8.12 Schematic representation of the four-cysteine zinc finger. Regions labelled A and B are of critical importance in determining, respectively, the DNA sequence that is bound by the finger and the optimal spacing between the two halves of the palindromic sequence that is recognized.

Table 8.2 Transcriptional regulatory proteins with multiple cysteine fingers

Finger type	Factor	Species
Cys_4—Cys_5	Steroid, thyroid receptors	Mammals
Cys_4	E1A	Advenovirus
Cys_6	GAL4, PPRI, LAC9	Yeast

capable of forming a pair of fingers, each with four cysteines coordinating a single zinc atom (Figure 8.12) and, as with the cysteine–histidine finger proteins, DNA binding of the receptors is dependent on the presence of zinc (Sabbah *et al.*, 1987).

However, the multi-cysteine finger cannot be converted into a functional cysteine–histidine finger by substituting two of its cysteine residues with histidines, suggesting that the two types of finger are functionally distinct (Green and Chambon, 1987). Moreover, unlike the cysteine–histidine zinc finger, which is present in multiple copies within the proteins that contain it, the unit of two multi-cysteine fingers present in the steroid receptors is found only once in each receptor. Interestingly, structural studies of the two multi-cysteine fingers in the glucocorticoid and oestrogen receptors (review: Schwabe and Rhodes, 1991) have indicated that the two fingers form one single structural motif consisting of two α-helices perpendicular to one another, with the cysteine–zinc linkage holding the base of a loop at the N-terminus of each helix (Figure 8.13). This is quite distinct from the modular structure of the two-cysteine two-histidine finger, where each finger constitutes an independent structural element whose configuration is unaffected by the presence or absence of adjacent fingers (Neuhaus *et al.*, 1990).

Thus, although these two DNA-binding motifs are similar in their coordination of zinc, they differ in the lack of histidines and of the conserved phenylalanine and leucine residues in the multi-cysteine finger, as well as structurally. It is clear, therefore, that they represent

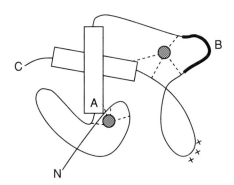

Figure 8.13 Schematic model of a pair of zinc fingers in the oestrogen receptor. Note the helical regions (indicated as cylinders) with the critical residues for determining the DNA sequence that is bound located at the terminus of the recognition helix (A), the zinc atoms (shaded), conserved basic residues (+++) and the region that interacts with another receptor molecule and determines the optimal spacing between the two halves of the palindromic sequence which is recognized (B). Note that A and B indicate the same regions as in Figure 8.12.

distinct functional elements and are unlikely to be evolutionarily related (review: Schwabe and Rhodes, 1991; Rhodes and Klug, 1993). Whatever the precise relationship between these motifs, it is clear that the multi-cysteine finger mediates DNA binding of the steroid–thyroid receptors. Thus mutations that eliminate or alter critical amino acids in this motif interfere with DNA binding by the receptor (Giguere *et al.*, 1986; Hollenberg *et al.*, 1987; Figure 8.14). The role of the cysteine fingers in mediating DNA binding by the steroid–thyroid hormone receptors can also be demonstrated by taking advantage of the observation (discussed in Section 4.3.1) that the different receptors bind to distinct but related sequences in the DNA of hormone-responsive genes (see Beato 1989 for review and Table 4.2 for a comparison of these binding sites). Thus, if the cysteine-rich region of the oestrogen receptor is replaced by that of the glucocorticoid receptor, the resulting chimaeric receptor has the DNA-binding specificity of the glucocorticoid receptor but continues to bind oestrogen, since all the other regions of the molecule are derived from the oestrogen receptor (Green and Chambon, 1987; Figure 8.15). Hence the DNA-binding specificity of the hybrid receptor is determined by its cysteine-rich region, resulting in

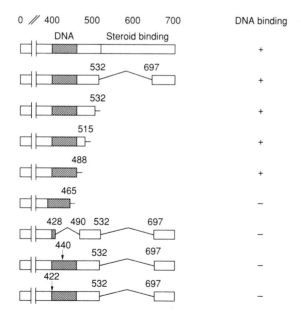

Figure 8.14 Effect of various deletions or mutations on the DNA binding of the glucocorticoid receptor. Note that DNA binding is prevented only by deletions that include part of the DNA-binding domain (shaded) or by mutations within it (arrows) but not by deletions in other regions such as the steroid-binding domain. Numbers indicate amino acid residues.

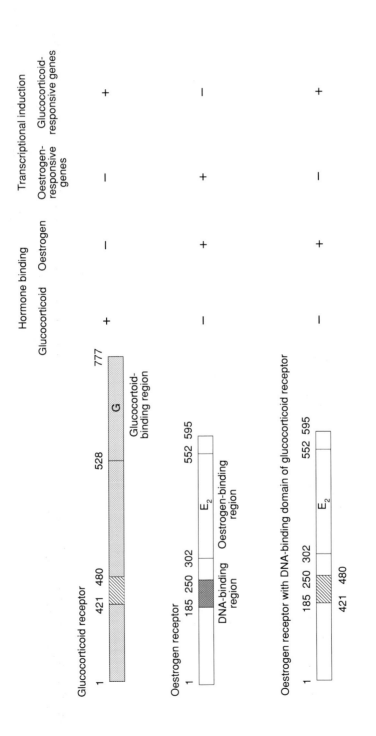

Figure 8.15 Effect of exchanging the DNA-binding domain (shaded) of the oestrogen receptor with that of the glucocorticoid receptor on the binding of hormone and gene induction by the hybrid receptor.

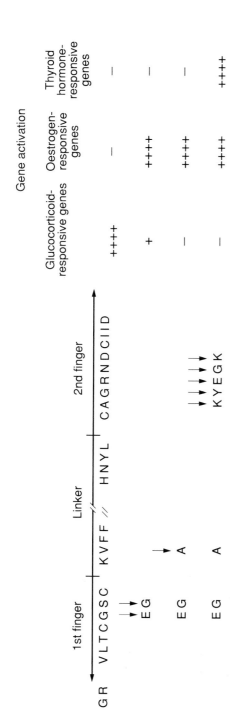

Figure 8.16 Effect of amino acid substitutions in the zinc finger region of the glucocorticoid receptor on the ability to bind to and activate genes that are normally responsive to different steroid hormones.

the hybrid receptor inducing the expression of glucocorticoid-responsive genes (which carry its DNA-binding site) in response to oestrogen (to which it binds).

These so-called 'finger-swap' experiments therefore provide further evidence in favour of the critical role for the multi-cysteine fingers in DNA binding, exchanging the fingers of two receptors exchanging the DNA-binding specificity. In addition, however, because of the existence of short distinct DNA-binding regions of this type in receptors which bind to distinct but related DNA sequences, they provide a unique opportunity to dissect the elements in a DNA-binding structure that mediate binding to specific sequences (review: Berg, 1989).

Thus, by exchanging one or more amino acids between two different receptors, it is possible to investigate the effects of these changes on DNA-binding specificity and hence elucidate the role of individual amino acid differences in producing the different patterns of sequence-specific binding. For example, the alteration of the two amino acids between the third and fourth cysteines of the N-terminal finger in the glucocorticoid receptor for their equivalents in the oestrogen receptor changes the DNA-binding specificity of the chimaeric receptor to that of the oestrogen receptor (Danielsen *et al.*, 1989; Umesono and Evans, 1989; Figure 8.16). Hence the exchange of two amino acids in a critical region of a protein of 777 amino acids (indicated as A in Figure 8.12) can completely change the DNA-binding specificity of the glucocorticoid receptor, resulting in it binding to and activating genes that are normally oestrogen responsive. The specificity of this hybrid receptor for such oestrogen-responsive genes can be further enhanced by exchanging another amino acid located between the two fingers (Figure 8.16), indicating that this region also plays a role in controlling the specificity of DNA binding.

As discussed in Section 4.3.1, the steroid receptors bind to palindromic recognition sequences within DNA, with the receptor binding to DNA as a dimer in which each receptor molecule interacts with one half of the palindrome. In addition to differences in the actual sequence recognized, steroid–thyroid hormone receptors can also differ in the optimal spacing between the two separate halves of the palindromic DNA sequence that is recognized (Table 4.2). Thus the oestrogen receptor and the thyroid hormone receptor both recognize the identical palindromic sequence in the DNA but differ in that in the thyroid receptor-binding sites the two halves of the palindrome are adjacent whereas in the oestrogen receptor-binding sites they are separated by three extra bases. The further alteration of the chimaeric receptor illustrated in Figure 8.16, the change of five amino acids in the second finger to their thyroid hormone receptor equivalents, is sufficient to allow the receptor to recognize thyroid hormone receptor-binding sites

(Figure 8.16). These amino acids in the second finger (indicated as B in Figure 8.12) appear to play a critical role, therefore, in determining the optimal spacing of the palindromic sequence that is recognized.

As discussed above, structural studies of the two zinc fingers in the oestrogen and glucocorticod receptors suggest that they form a single structural motif with two perpendicular α-helices (Figure 8.13). In this proposed structure, the critical amino acids for determining the spacing in the palindromic sequence recognized are located on the surface of the molecule, allowing them to interact with equivalent residues on another receptor monomer during dimerization (indicated as B in Figure 8.17). Hence differences in the interaction of these regions in the different receptors determine the spacing of the two monomers within the receptor dimer and thus the optimal spacing in the palindromic DNA sequence that is recognized.

Interestingly, within this structure, the critical residues for determining the precise DNA sequence that is recognized are located at the N-terminus of the first α-helix (indicated as A in Figure 8.17), further supporting the critical role of such helices in DNA binding. Moreover, in the proposed structure of the oestrogen receptor dimer, the DNA-binding helices in each monomer will be separated by 34Å, allowing each of these recognition helices to make sequence-specific contacts in adjacent major grooves of the DNA molecule.

The definition of the DNA-binding domain of the steroid–thyroid hormone receptors as a short sequence containing two multi-cysteine fingers has therefore allowed the elucidation of the features in this motif that mediate the different sequence specificities of the different recep-

Figure 8.17 Interaction of two oestrogen receptor molecules to form a DNA-binding dimer. Compare with Figure 8.13 and note the interaction of the B regions on each molecule. The resulting dimer has a spacing of 34 Å between the two DNA-binding regions, allowing binding in successive major grooves of the DNA molecule.

tors and their relationship to the structure of the motif. In particular, a helical region of the first finger plays a critical role in determining the precise DNA sequence that is recognized by binding in the major groove of the DNA, whilst a distinct region in the second zinc finger controls the spacing of adjacent sequences which is optimal for the binding of each receptor by interacting with the equivalent region in another receptor monomer and hence affecting the structure of the receptor dimer that forms.

8.4 THE LEUCINE ZIPPER AND THE BASIC DNA-BINDING DOMAIN

As discussed in the preceding sections of this chapter, the study of motifs common to several different transcription factors has led to the identification of the role of these motifs in DNA binding. A similar approach has led to the identification of the leucine zipper motif (reviews: Lamb and McKnight, 1991; Hurst, 1994; Kerppola and Curran, 1995). Thus this structure has been detected in several different transcription factors, such as the CCAAT box-binding protein C/EBP (see Section 3.3.3), the yeast factor GCN4 and the oncogene products Myc, Fos and Jun (see Sections 7.2.1 and 7.2.4). It consists of a leucine-rich region in which successive leucine residues occur every seventh amino acid (Figure 8.18).

In all these cases, the leucine-rich region can be drawn as an α-helical structure in which adjacent leucine residues occur every two turns on the same side of the helix. Moreover, these leucine residues appear to play a critical role in the functioning of the protein. Thus, with one exception (a single methionine in the Myc protein), the central leucine residues of the motif are conserved in all the factors that contain it (Figure 8.18). It was therefore proposed (Landschulz *et al.*, 1988) that the long side-chains of the leucine residues extending from one polypep-

C/EBP	L	T S D N D R	L	R K R V E Q	L	S R E L D T	L	R G I F R Q	L
Jun B	L	E D K V K T	L	K A E N A G	L	S S A A G L	L	R E Q V A Q	L
Jun	L	E E K V K T	L	K A Q N S E	L	A S T A N M	L	R E Q V A Q	L
GCN 4	L	E D K V E E	L	L S K N Y H	L	E H E V A R	L	K K L V G E R
Fos	L	Q A E T D Q	L	E D E K S A	L	Q T E I A N	L	L K E K E K	L
Fra 1	L	Q A E T D K	L	E D E K S G	L	Q R E I I E	L	Q K Q K E R	L
c-Myc	V	Q A E E Q K	L	I S E E D L	L	R K R R E Q	L	K H K L E Q	L
n-Myc	L	Q A E E H Q	L	L L E K E K	L	Q A R Q Q Q	L	L K K I E H A
l-Myc	L	V G A E K K M	A T E K R Q	L	R C R Q Q Q	L	Q K R I A Y	L

Figure 8.18 Alignment of the leucine-rich region in several cellular transcription factors. Note the conserved leucine residues (L) which occur every seven amino acids.

Figure 8.19 Model of the leucine zipper and its role in the dimerization of two molecules of a transcription factor.

tide would interdigitate with those of the analogous helix of a second polypeptide, forming a motif known as the leucine zipper, which would result in the dimerization of the factor (Figure 8.19). This effect could also be achieved by a methionine residue which, like leucine, has a long side-chain with no lateral methyl groups, but not by other hydrophopic amino acids such as valine or isoleucine which have methyl groups extending laterally from the β-carbon atom.

In agreement with this idea, substitutions of individual leucine residues in C/EBP or other leucine zipper-containing proteins such as Myc, Fos and Jun with isoleucine or valine abolish the ability of the intact protein to form a dimer, indicating the critical role of this region in dimerization (Landschulz *et al.*, 1989; review: Johnson and McKnight, 1989). A comparison of the effects of various mutations of this type on the ability of the mutant protein to dimerize suggested that the two leucine-rich regions associate in a parallel manner, with both helices oriented in the same direction (as illustrated in Figure 8.19) rather than in an anti-parallel configuration as originally suggested (Landschulz *et al.*, 1989). This idea was confirmed by structural studies of the leucine zipper regions of GCN4 and in the Fos–Jun dimer bound to DNA (Glover and Harrison, 1995). These studies indicated that each zipper motif forms a right-handed α-helix, with dimerization occurring via the association of two parallel helices that coil around each other to form a coiled coil motif similar to that found in fibrous proteins such as the keratins and myosins (Figure 8.20).

In addition to its role in dimerization, the leucine zipper is also essential for DNA binding by the intact molecule. Thus mutations in the zipper which prevent dimerization also prevent DNA binding from occurring (Landschulz *et al.*, 1989). Unlike the zinc finger or helix–turn–helix motifs, however, the zipper is not itself the DNA-binding domain of the molecule and does not directly contact the DNA. Rather, it

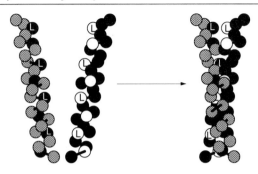

Figure 8.20 Coiled coil structure of the leucine zipper formed by two helical coils wrapping around each other. L indicates a leucine residue.

facilitates DNA binding by an adjacent region of the molecule which, in C/EBP, Fos and Jun, is rich in basic amino acids and can therefore interact directly with the acidic DNA. The leucine zipper is believed, therefore, to serve an indirect structural role in DNA binding, facilitating dimerization, which in turn results in the correct positioning of the two basic DNA-binding domains in the dimeric molecule for DNA binding to occur (Figure 8.21).

In agreement with this idea, mutations in the basic domain abolish the ability to bind to DNA without affecting the ability of the protein to dimerize, as expected for mutations that affect the DNA binding domain directly (Landschulz *et al.*, 1989). Similarly, exchange of the basic region of GCN4 for that of C/EBP results in a hybrid protein with the DNA-binding specificity of C/EBP, whilst exchange of the leucine zipper region has no effect on the DNA-binding specificity of the hybrid molecule (Agre *et al.*, 1989; Figure 8.22).

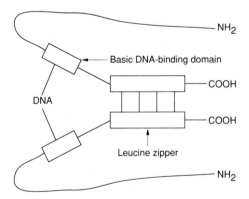

Figure 8.21 Model for the structure of the leucine zipper and the adjacent DNA-binding domain following dimerization of the transcription factor C/EBP.

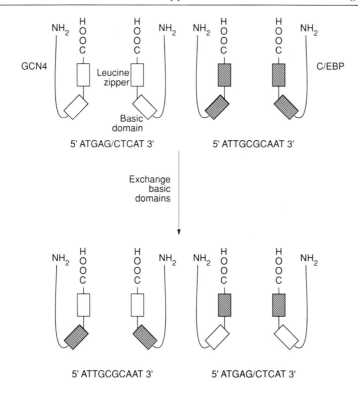

Figure 8.22 Effect of exchanging the basic domains of GCN4 and C/EBP on the DNA-binding specificity. Note that the DNA-binding specificity is determined by the origin of the basic domain and not that of the leucine zipper.

Hence the DNA-binding specificity of leucine zipper-containing transcription factors is determined by the sequence of their basic domain, with the leucine zipper allowing dimerization to occur and hence facilitating DNA binding by the basic domain. As expected from this idea, the basic DNA-binding domain can interact with DNA in a sequence-specific manner in the absence of the leucine zipper if it is first dimerized via an intermolecular disulphide bond (Talanian *et al.*, 1990; Figure 8.23). Interestingly, the basic DNA-binding domain can bind to DNA as a monomer in the case of the Skn-1 factor, which lacks a leucine zipper (Blackwell *et al.*, 1994). In this factor, however, the basic domain is part of a composite DNA-binding domain which also contains a region homologous to the N-terminal arm of the homeobox (see Section 8.2).

In factors having a simple basic DNA-binding domain, following dimerization via the leucine zipper, the intact transcription factor will form a rotationally symmetrical dimer that contacts the DNA via the bifurcating basic regions (Figure 8.21) which form α-helical structures.

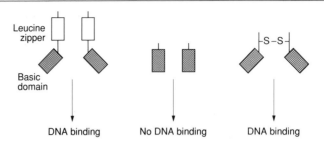

Figure 8.23 DNA binding of molecules containing basic DNA-binding domains can occur following dimerization mediated by leucine zippers or by a disulphide bridge (S–S) but cannot be achieved by unlinked monomeric molecules.

These two helices then track along the DNA in opposite directions corresponding to the dyad symmetrical structure of the DNA recognition site and form a clamp or scissors grip around the DNA, similar to the grip of a wrestler on his or her opponent, resulting in very tight binding of the protein to DNA (Vinson *et al.*, 1989; Glover and Harrison, 1995). Most interestingly, structural studies have suggested that the basic region does not assume a fully α-helical structure until it contacts the DNA, when it undergoes a configurational change to a fully α-helical form. Hence the association of the transcription factor with the appropriate DNA sequence results in a conformational change in the factor, leading to a tight association with that sequence (for discussion see Sauer (1990)).

Although originally identified in the leucine zipper-containing proteins, the basic DNA-binding domain has also been identified by a homology comparison in a number of other transcription factors that do not contain a leucine zipper (Prendergast and Ziff, 1989). These factors include the immunoglobulin enhancer-binding proteins, E12 and E47, discussed in Sections 5.2 and 5.3.2, the muscle-determining gene *MyoD* (Section 5.3) and the *Drosophila* daughterless protein.

In these cases, the basic DNA-binding domain is juxtaposed to a region that can form a helix–loop–helix motif (review: Littlewood and Evan, 1994). This helix–loop–helix motif is distinct from the helix–turn–helix motif in the homeobox (Section 8.2) in that it can form two amphipathic helices, containing all the charged amino acids on one side of the helix, which are separated by a non-helical loop (Murre *et al.*, 1989a). This helix–loop–helix motif is thought to play a similar role to the leucine zipper, allowing dimerization of the transcription factor molecule and thereby facilitating DNA binding by the basic motif (Murre *et al.*, 1989b; for discussion see Jones (1990)).

In agreement with this, deletion or mutations in the basic domain of the MyoD protein do not abolish dimerization but do prevent DNA

		Dimerization	DNA binding	Muscle-specific gene activation
Intact MyoD	Basic | Helix | Loop | Helix	+	+	+
		+	−	−
		+	−	−
		−	−	−
		−	−	−
		−	−	−
	E12	+	+	−

Figure 8.24 Effect of deleting the basic domain or the adjacent helix–loop–helix motif on dimerization, DNA binding and activation of muscle-specific gene expression by the MyoD transcription factor. Note that deletion of any part of the helix–loop–helix motif abolishes dimerization and consequent DNA binding and gene activation, whilst deletion of the basic domain directly abolishes DNA binding and consequent gene activation. Substitution of the basic domain of the constitutive factor E12 for that of MyoD restores DNA binding but not the ability to activate muscle-specific gene expression.

binding, paralleling the effect of similar mutations in C/EBP (Figure 8.24). Similarly, mutations or deletions in the helix–loop–helix region abolish both dimerization and DNA binding, parallelling the effects of similar mutations in leucine zipper-containing proteins. Moreover, the DNA binding ability of MyoD from which the basic DNA-binding domain has been deleted can be restored by substituting the basic domain of the E12 protein (Davis *et al.*, 1990). However, such substitution does not allow the hybrid protein to activate muscle-specific gene expression, suggesting that, in addition to mediating DNA binding, the basic region of MyoD also contains elements involved in the activation of muscle-specific genes (Davis *et al.*, 1990; Figure 8.24).

Interestingly, it has been shown that the conversion of three amino acids within the E12 basic region to their MyoD equivalents allows the E12 basic region to activate muscle-specific gene expression following DNA binding (Figure 8.25) (Davis and Weintraub, 1992). The recently observed crystal structure of MyoD bound to DNA (Ma *et al.*, 1994) suggests that these amino acids may play a critical role in allowing the MyoD basic region to assume a particular structural configuration in which it can interact with other activating transcription factors. In agreement with this idea, the substitution of these same three amino acids in E12 for their MyoD equivalents allows the mutant E12 protein to bind to another muscle-specific transcription factor MEF2A, which is normally a property of MyoD alone (Figure 8.25) (Kanshal *et al.*, 1994).

	DNA binding	Muscle-specific gene activation	Interaction with MEF2A
MyoD ——A T——K—— (114 115 ... 124)	+	+	+
E12 ----N N----D----	+	−	−
E12(M) ----A T----K----	+	+	+

Figure 8.25 Alterations of three amino acids (positions 114, 115 and 124) in the E12 basic domain to their MyoD equivalents confers on the resulting protein (E12(M)) the ability to interact with the muscle-specific transcription factor MEF2A and activate muscle-specific genes following DNA binding, which are normally properties of MyoD alone.

Hence, like the POU domain (see Section 6.3.2), the basic domain appears to function both as a DNA-binding domain and as a site for protein–protein interactions critical for transcriptional activation.

Both the leucine zipper and the helix–loop–helix motif therefore act by causing dimerization, allowing DNA binding by the adjacent basic motif. Interestingly, the Myc oncoproteins contain both a helix–loop–helix motif and a leucine zipper region adjacent to the basic DNA-binding region (Landschulz *et al.*, 1988; Murre *et al.*, 1989a). Moreover, the leucine zipper can also be found as a dimerization motif in proteins that use DNA-binding motifs other than the basic region. Thus, in the *Arabidopsis* Athb-1 and 2 proteins, the leucine zipper facilitates dimerization, with DNA binding being produced by the adjacent homeobox (Sessa *et al.*, 1993). Thus individual DNA binding and dimerization motifs can be combined in different combinations to produce molecules capable of dimerizing and binding to DNA.

The essential role of dimerization (mediated by the leucine zipper or the helix–loop–helix motifs) in allowing DNA binding by basic DNA-binding domain proteins provides an additional aspect to the regulation of these factors (for discussion see Jones (1990) and Lamb and McKnight (1991)). Thus, in addition to the formation of homodimers, it is possible to hypothesize that heterodimers will also form between two different leucine zipper- or two different helix–loop–helix-containing factors, allowing the production of dimeric factors with novel DNA-binding specificities or affinities for different sites. One example of this process is seen in the oncogene products Fos and Jun. Thus, as discussed in Section 7.2.1, the Fos protein cannot bind to AP1 sites in DNA when present alone but can form a heterodimer with the Jun protein which is capable of binding to such sites with 30-fold greater affinity than a Jun homodimer (Figure 8.26). The formation of Jun homodimers and Jun–Fos heterodimers is dependent upon the leucine zipper regions of the proteins. Moreover, the failure of Fos to form homodimers is similarly

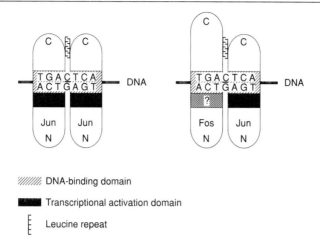

Figure 8.26 Model for DNA binding by the Jun homodimer and the Fos–Jun heterodimer.

dependent on its leucine zipper region. Thus, if the leucine zipper domain of Fos is replaced by that of Jun, the resulting protein can dimerize (Neuberg *et al.*, 1989) and the chimaeric protein can bind to DNA through the basic DNA-binding region of Fos, which is therefore a fully functional DNA-binding domain. Hence the ability of leucine zipper proteins to bind to DNA is determined both by the nature of the leucine zipper which facilitates homodimerization and/or heterodimerization as well as by the basic DNA-binding motif which allows DNA binding following dimerization (for discussion see Kerppola and Curran (1995)).

In addition to its positive role in allowing DNA binding by factors that cannot do so as homodimers, heterodimerization between two related factors can also have an inhibitory role. Thus, as discussed in Section 5.3.2, the DNA-binding ability of functional helix–loop–helix proteins which contain a basic DNA-binding domain can be inhibited by association with the Id protein. This protein contains a helix–loop–helix motif, allowing it to associate with other members of this family, but lacks the basic DNA-binding domain. The heterodimer of Id and a functional protein therefore lacks the dimeric basic regions necessary for DNA binding, and the activity of the functional transcription factor is thereby inhibited by Id.

Hence the role of the leucine zipper and helix–loop–helix motifs in dimerization can be put to use in gene regulation in both positive and negative ways, either allowing DNA binding by factors that could not do so in isolation or inhibiting the binding of fully functional factors.

8.5 OTHER DNA-BINDING MOTIFS

Although the majority of DNA-binding domains that have been identified in known transcription factors fall into the three classes discussed in the preceeding sections, not all do so. Thus, for example, the DNA-binding domains of transcription factors such as AP2, and the CCAAT box-binding factors CTF/NF1 and HAP2 and HAP3 (see Section 3.3.2) are distinct from the known motifs and from each other. As more and more factors are cloned, it is likely that other factors with DNA-binding motifs similar to those of these proteins will be identified and that they will become founder members of new families of DNA-binding motifs. Indeed, this process is already underway, the relatedness of the DNA-binding domains of the *Drosophila* transcription factor fork head and the mammalian liver transcription factor HNF-3 having led to the identification of a new family of proteins containing the so-called fork head DNA-binding motif (Lai *et al.*, 1993), whilst another recently identified domain common to the *ets*-1 proto-oncogene, the *Drosophila* E74 gene, the mouse PU-1 gene and several other transcription factors has been identified as a new DNA-binding motif mediating binding to the purine-rich sequences recognized by these factors (Karim *et al.*, 1990). Similarly, the ribosomal RNA transcription factor, upstream binding factor (UBF) (see Section 3.2.2), contains a DNA-binding domain that has also been identified in several other factors, including high-mobility group (HMG) proteins, and which is therefore known as the HMG box (Grosschedel *et al.*, 1994), whilst the DNA-binding domain in the p53 protein (discussed in Section 7.3.2) has been shown to be related to that of the NFκB family (discussed in Section 5.2.2) (Muller *et al.*, 1995; review: Baltimore and Beg, 1995).

8.6 CONCLUSIONS

In this chapter we have discussed a number of different DNA-binding motifs common to several different transcription factors that can mediate DNA binding. These motifs are listed in Table 8.3.

As we have seen, it has proved possible in many cases to define the precise amino acids in a particular motif which mediate binding to a particular DNA sequence. It should be noted, however, that many transcription factors have the ability to bind to several dissimilar sequences using the same DNA-binding domain. For example, as discussed in Chapter 4, the glucocorticoid receptor binds to a specific DNA

Table 8.3 DNA-binding motifs

Motif	Structure	Factors containing domain	Comments
Homeobox	Helix–turn–helix	Numerous *Drosophila* homeotic genes, related genes in other organisms	Structurally related to similar motif in bacteriophage proteins
POU	Helix–turn–helix	Mammalian Oct-1, Oct-2, Pit-1, nematode *unc86*	Related to homeodomain
Cysteine–histidine zinc finger	Multiple fingers each coordinating a zinc atom	TFIIIA, Kruppel, Sp1, etc.	May form β-sheet and adjacent α-helical structure
Cysteine–cysteine zinc finger	Single pair of fingers each coordinating a zinc atom	Steroid–thyroid hormone receptor family	Related motifs in EIA, GAL4, etc.
Basic domain	α-helical	C/EBP c-Fos, c-Jun, c-Myc, MyoD, etc	Associated with leucine zipper and/or helix–loop–helix dimerization motifs
Fork head	Unknown	Fork head, HNF 3A	—
ETS	Unknown	c-Ets, c-Erg, *Drosophila* E74, PU. 1	Binds purine-rich sequences

sequence in genes that are induced by glucocorticoid and to a distinct DNA sequence in genes that are repressed by this hormone, the same DNA-binding domain of the protein being used in each case (Sakai *et al.*, 1988). Similarly, the yeast *CYC1* and *CYC7* genes contain entirely distinct sequences, both of which bind the HAP1 transcription factor, allowing gene activation to occur (Pfeifer *et al.*, 1987).

This phenomenon of a single factor binding to highly divergent sequences has been most extensively analysed in the case of the octa-mer-binding protein Oct-1. Thus this factor binds to a sequence in the SV40 enhancer which shares less than 30% homology (4 of 14 bases), or little more than a random match, with another Oct-1-binding sequence in the herpes simplex virus immediate-early gene promoters (Figure 8.27). By analysing a series of other Oct-1-binding elements, however, Baumruker *et al.* (1988) were able to show that the two apparently unrelated Oct-1-binding sites could be linked by a smooth progression via a series of other binding sites related to one another (Figure 8.27). This suggests, therefore, that Oct-1 can bind to very dissimilar sequences because there are few, if any, obligatory contacts with specific bases in potential binding sites. Rather, specific binding to a particular sequence can occur via many possible independent interactions with DNA, only some of which will occur with any particular binding site. Hence the binding to apparently unrelated sequences does not reflect two distinct binding specificities but indicates that the protein can make many different contacts with DNA, the sequences that can specifically bind the protein being those with which it can make a certain proportion of these possible contacts.

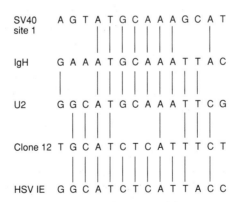

Figure 8.27 Relationship between the various diverse sequences bound by the Oct-1 transcription factor in the simian virus 40 enhancer, the immunoglobulin H chain gene enhancer (IgH), the U2 snRNA gene, clone 12 (a mutated version of a site in the SV40 enhancer which binds Oct-1) and the herpes simplex virus immediate-early genes (HSV IE).

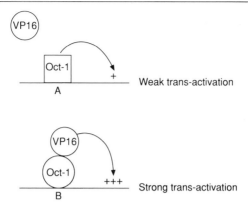

Figure 8.28 The octamer-binding protein Oct-1 binds to most binding sites (A) in a configuration that is not recognized by VP16. This results in only the weak trans-activation characteristic of Oct-1 alone. In contrast, when it binds to its binding sites in the HSV IE promoters (B), Oct-1 undergoes a conformational change allowing it to be recognized by the strong trans-activator VP16, leading to strong trans-activation.

Interestingly, it has been shown that the secondary structure of Oct-1 bound to these sites differs, so that its configuration when bound to the HSV IE sequence is different to that observed when it is bound to the other sequences (Walker *et al.*, 1994). Moreover, this configurational change allows the Oct-1 bound to the HSV promoter to be recognized by the HSV VP16 protein, whereas this does not occur with Oct-1 bound to other sequences. As discussed in Section 9.2.1, VP16 is a much stronger trans-activator than Oct-1 alone, and this therefore results in the strong activation of the HSV IE promoters by the Oct-1–VP16 complex, whereas other promoters in which Oct-1 has bound to different sequences are insensitive to such trans-activation by VP16. Hence this provides a novel example of gene regulation in which the nature of the sequence bound by a factor controls its recognition by another factor, resulting in strong trans-activation only from a subset of sequences bound by Oct-1 (Figure 8.28).

In addition to the ability of one factor to bind to different sequences and have different effects, it is also possible for the same sequence to be bound by more than one factor. Thus, as well as the Oct-1 factor, several other proteins bind to the octamer motif, including the B cell-specific factor Oct-2 (see Section 5.2.2) and several others expressed specifically in the brain (Scholer *et al.*, 1989). In all these factors, DNA binding is mediated by the POU domain discussed above (Section 8.2). In contrast, however, whilst the transcription factors CTF/NF1 and C/EBP both bind to the CCAAT box sequence (see Section 3.3.3), they do so via completely different DNA-binding domains with C/EBP having a basic DNA-binding domain (Section 8.4) whilst CTF/NF1 has a DNA-binding domain distinct from that of any other factor (Section 8.5).

It is unlikely, therefore, that the existence of several distinct DNA-binding domains reflects the need of the factors that contain them to bind to distinct types of DNA sequences. Rather, it seems perfectly possible that one DNA-binding motif could be present in all factors, with variations of it in different factors producing the observed binding to different DNA sequences. This is particularly so in view of the fact that in diverse DNA-binding motifs, such as the homeobox, the basic DNA-binding domain and the two types of zinc fingers, the amino acids that determine sequence-specific binding to DNA are all located within similar α-helical structures. This idea evidently begs the question of why different DNA-binding motifs exist.

It is possible that this situation has arisen simply by different motifs that could produce DNA binding having arisen in particular factors during evolution and having been retained since they efficiently fulfilled their function. Alternatively, it may be that the existence of different motifs reflects differences in the factors containing them other than the specific DNA sequence that is recognized. For example, the highly repeated zinc finger motif may be of particular use where, as in the case of transcription factor TFIIIA, the factor must contact a large regulatory region in the DNA. Similarly, a motif such as the basic domain, which can bind to DNA only following dimerization, will be of particular use where the activity of the factor must be regulated, whether positively or negatively, via dimerization with another factor.

Whatever the case, it is clear that DNA binding by transcription factors is dependent upon specific domains of defined structure within the molecule. Following such DNA binding, the bound factor must influence the rate of transcription, either positively or negatively. The manner in which this occurs and the regions of the factors that achieve this effect are discussed in chapter 9.

REFERENCES

Agre, P., Johnson, P. F. and McKnight, S. L. (1989). Cognate DNA binding specificity retained after leucine zipper exchange between GCN4 and C/EBP. *Science* **246**, 922–926.

Baltimore, D. and Beg, A. E. (1995). DNA-binding proteins: a butterfly flutters by. *Nature* **373**, 287–288.

Baumruker, T., Sturm, R. and Herr, W. (1988). OBP 100 binds remarkably degenerate octamer motifs through specific interaction with flanking sequences. *Genes and Development* **2**, 1400–1413.

Beato, M. (1989). Gene regulation by steroid hormones. *Cell* **56**, 335–344.

Berg, J. M. (1988). Proposed structure for the zinc-binding domains from transcription factor III A and related proteins. *Proceedings of the National Academy of Sciences USA* **85**, 99–102.

Berg, J. M. (1989). DNA binding specificity of steroid receptors. *Cell* **57**, 1065–1068.

Blackwell, T. K., Bowerman, B., Priess, J. R. and Weintraub, H. (1994). Formation of a monomeric DNA binding domain by Skn-1 bZip and homeodomain elements. *Science* **266**, 621–628.

Choo, Y. and Klug, A. (1994). Toward a code for the interaction of zinc fingers with DNA: selection of randomized fingers displayed on phage. *Proceedings of the National Academy of Sciences USA* **91**, 11163–11167.

Danielsen, M., Hinck, C. and Ringold, G. M. (1989). Two amino acids within the knuckle of the first zinc finger specify DNA response element activation by the glucocorticoid receptor. *Cell* **57**, 1131–1138.

Davis, R. L. and Weintraub, H. (1992). Acquisition of myogenic specificity by replacement of three amino acid residues from MyoD into E12. *Science* **256**, 1027–1030.

Davis, R. L., Cheng, P-F., Lassar, A. B. and Weintraub, H. (1990). The MyoD DNA binding domain contains a recognition code for muscle-specific gene activation. *Cell* **60**, 733–746.

Diakun, G. P., Gairall, L. and Klug, A. (1986). EXAFS study of the zinc-binding sites in the protein transcription factor III A. *Nature* **324**, 698–699.

Evans, R. M. and Hollenberg, S. M. (1988). Zinc fingers: guilt by association. *Cell* **52**, 1–3.

Freedman, L. P., Luisi, B. F., Korszin, Z. R., Basavappa, R., Sigler, P. B. and Yamamoto, K. R. (1988). The function and structure of the metal coordination sites within the glucocorticoid receptor binding domain. *Nature* **334**, 543–546.

Gehring, W. J., Qian, Y. Q., Billeter, M., Furukubo-Tukunagu, K., Schier, A. F., Resendez-Perez, D., Affolter, M., Otting, G. and Wuthrich, K. (1994). Homeodomain-DNA recognition. *Cell* **78**, 211–223.

Giguere, V., Hollenberg, S. M., Rosenfeld, M. G. and Evans, R. M. (1986). Functional domains of the human glucocorticoid receptor. *Cell* **46**, 645–652.

Glover, J. N. M. and Harrison, S. C. (1995). Crystal structure of the heterodimeric bZip transcription factor c-Fos-c-Jun bound to DNA. *Nature* **373**, 257–261.

Green, S. and Chambon, P. (1987). Oestradiol induction of a glucocorticoid-response gene by a chimaeric receptor. *Nature* **325**, 75–78.

Grosschedel, R., Giese, K. and Pagel, J. (1994). HMG proteins: architectural elements in the assembly of nucleoprotein structures. *Trends in Genetics* **10**, 94–100.

Hanas, J. S., Bogenhagen, D. F. and Wu, C-W. (1983). Co-operative model for the binding of *Xenopus* transcription factor A to the 5*S* RNA gene. *Proceedings of the National Academy of Sciences USA* **80**, 2142–2145.

Harrison, S. C. (1991). A structural taxonomy of DNA binding domains. *Nature* **353**, 715–719.

Hollenberg, S. M., Giguere, V., Segui, P. and Evans, R. M. (1987). Colocalization of DNA-binding and transcriptional activation functions in the human glucocortidoid receptor. *Cell* **49**, 39–46.

Hurst, H. C. (1994). bZIP proteins. *Protein Profile* **1**, 123–168.

Johnson, P. F. and McKnight, S. L. (1989). Eukaryotic transcriptional regulatory proteins. *Annual Review of Biochemistry* **58**, 799–839.

Jones, N. (1990). Transcriptional regulation by dimerization: two sides to an incestuous relationship. *Cell* **61**, 9–11.

Kadonaga, J. T., Carner, K. R., Masiarz, F. R. and Tjian, R. (1987). Isolation of cDNA encoding the transcription factor Sp1 and functional analysis of the DNA binding domain. *Cell* **51**, 1079–1090.

Kanshal, S., Schneider, J. W., Nudal-Ginard, B. and Mahdavi, V. (1994). Activation of the myogenic lineage by MEF2A, a factor that induces and cooperates with MyoD. *Science* **266**, 1236–1240.

Karim, F. D., Urness, L. D., Thummel, C. S., Klemsz, M. J., McKercher, S. R., Celada, A., Van Beveren, C., Maki, R. A., Gunther, C. V., Nye, J. A. and Graves, B. J. (1990). The Ets-domain: a new DNA-binding motif that recognizes a purine-rich core DNA sequence. *Genes and Development* **4**, 1451–1453.

Kerppola, T. and Curran, T. (1995). Zen and the art of Fos and Jun. *Nature* **373**, 199–200.

Klemm, J. D., Rould, M. A., Aurora, R., Herr, W. and Pabo, C. O. (1994). Crystal structure of the Oct-1 POU domain bound to an octamer site: DNA recognition with tethered DNA-binding molecules. *Cell* **77**, 21–23.

Klug, A. and Rhodes, D. (1987). Zinc fingers: a novel protein motif for nucleic acid recognition. *Trends in Biochemical Sciences* **12**, 464–469.

Kornberg, T. B. (1993). Understanding the homeodomain. *Journal of Biological Chemistry* **268**, 26813–26816.

Lai, E., Clark, K. L., Burley, S. K. and Darnell, J. E. Jr. (1993). Hepatocyte nuclear factor 3/fork head or 'winged helix' proteins: a family of transcription factors of diverse biological function. *Proceedings of the National Academy of Sciences USA* **90**, 10421–10423.

Lamb, P. and McKnight, S. L. (1991). Diversity and specificity in transcriptional regulation: the benefits of heterotypic dimerization. *Trends in Biochemical Sciences* **16**, 417–422.

Landschulz, W. H., Johnson, P. F. and McKnight, S. L. (1988). The leucine zipper: a hypothetical structure common to a new class of DNA binding proteins. *Science* **240**, 1759–1764.

Landschulz, W. H., Johnson, P. F. and McKnight, S. L. (1989). The DNA binding domain of the rat liver nuclear protein C/EBP is bipartite. *Science* **243**, 1681–1688.

Littlewood, T. and Evan, G. (1994). Helix–loop–helix. *Protein Profile* **1**, 639–709.

Ma, P. C. M., Rould, M. A., Weintraub, H and Pabo, C. O. (1994). Crystal structure of MyoD bHLH domain—DNA complex: perspectives on DNA recognition and implications for transcriptional activation. *Cell* **77**, 451–459.

Miller, J., McLachlan, A. D. and Klug, A. (1985). Repetitive zinc-binding domains in the protein transcription factor III A from *Xenopus* oocytes. *EMBO Journal* **4**, 1609–1614.

Muller, C. W., Rey, F. A., Sodeoka, M., Verdine, G. L. and Harrison, S. C. (1995). Structure of the NF-κB p50 homodimer bound to DNA. *Nature* **373**, 311–317.

Murre, C., McCaw, P. S. and Baltimore, D. (1989a). A new DNA binding and dimerization motif in immunoglobulin enhancer binding, daughterless, MyoD and myc proteins. *Cell* **56**, 777–783.

Murre, C., McCaw, P. S., Vaessin, H., Caudy, M., Jan, L. Y., Jan, Y. N., Cabera, C. V., Buskin, J. N., Hauschka, S. D., Lassar, A. B., Weintraub, H. and Baltimore, D. (1989b). Interactions between heterologous helix–loop–helix proteins generate complexes that bind specifically to a common DNA sequence. *Cell* **58**, 537–544.

Nardelli, J., Gibson, T. J., Vesque, C. and Charnay, P. (1991). Base sequence discrimination by zinc-finger DNA binding domains. *Nature* **349**, 175–178.

Neuberg, M., Adamkiewicz, J., Hunter, J. B. and Muller, R. (1989). Fos protein containing the Jun leucine zipper forms a homodimer which binds to the AP1 binding site. *Nature* **341**, 243–245.

Neuhaus, D., NaKaseko, Y., Nagai, A. and Klug, A. (1990). Sequence specific [^{1}H] NMR resonance assignments and secondary structure identification for 1- and 2-zinc finger constructs from SWI5. *FEBS Letters* **262**, 179–184.

Pabo, C. and Sauer, R. T. (1992). Transcription factors: structural families and principles of DNA recognition. *Annual Review of Biochemistry* **61**, 1053–1095.

Parker, M. G. (1993). Steroid and related receptors. *Current Opinion in Cell Biology* **5**, 499–504.

Parraga, G., Horvath, S. J., Eisen, A., Taylor, W. E., Hood, L., Young, E. T. and Klevitt, R. E. (1988). Zinc dependent structure of a single-finger domain of yeast ADRI. *Science* **241**, 1489–1492.

Pfeifer, K., Prezant, T. and Guarente, L. (1987). Yeast HAPI activator binds to two upstream sites of different sequence. *Cell* **49**, 19–27.

Prendergast, G. C. and Ziff, E. B. (1989). DNA-binding motif. *Nature* **341**, 392.

Rebar, E. J. and Pabo, C. O. (1994). Zinc finger phage: affinity selection of fingers with new DNA-binding specificities. *Science* **263**, 671–673.

Redemann, N., Gaul, U. and Jackle, H. (1988). Disruption of a putative Cys–zinc interaction eliminates the biological activity of the Kruppel finger protein. *Nature* **332**, 90–92.

Rhodes, D. and Klug, A. (1993). Zinc finger structure. *Scientific American* **268**, 32–39.

Rosenberg, U. B., Schroder, C., Preiss, A., Kienlin, A., Cote, S., Riede, I. and Jackle, H. (1986). Structural homology of the *Drosophila Kruppel* gene with *Xenopus* transcription factor III A. *Nature* **319**, 336–339.

Sabbah, M., Redeuilh, G., Secco, C. and Baulieu, E.-E. (1987). The binding activity of estrogen receptor DNA and heat shock protein (Mr 90,000) is dependent on receptor bound metal. *Journal of Biological Chemistry* **262**, 8631–8635.

Sakai, D. D., Helms, S., Carlstedt-Duke, J., Gustafsson, J-A., Rottman, F. M. and Yamamoto, K. R. (1988). Hormone-mediated repression: a negative glucocorticoid response element from the bovine prolactin gene. *Genes and Development* **2**, 1144–1154.

Sauer, R. T. (1990). Scissors and helical forks. *Nature* **347**, 514–515.

Scholer, H. R., Hatzpoulos, A. K., Balling, A. R., Suzuki, N. and Gruss, P. (1989). A family of octamer-specific proteins during mouse embryogenesis: evidence for germline-specific expression of an Oct factor. *EMBO Journal* **8**, 2543–2550.

Schwabe, J. W. R. and Rhodes, D. (1991). Beyond zinc fingers: steroid hormone receptors have a novel structural motif for DNA recongition. *Trends in Biochemical Sciences* **16**, 291–296.

Sessa, G., Morelli, G. and Ruberti, I. (1993). The Athb-1 and -2 HD-ZIP domains homodimerize forming complexes of different DNA binding specificities. *EMBO Journal* **12**, 3507–3517.

Struhl, K. (1989). Helix–turn–helix, zinc finger, and leucine zipper motifs for eukaryotic transcriptional regulatory proteins. *Trends in Biochemical Sciences* **14**, 137–140.

Talanian, R. V., McKnight, J. C. and Kim, P. S. (1990). Sequence-specific DNA binding by a short peptide dimer. *Science* **249**, 769–771.

Travers, A. (1993). *DNA Protein Interactions*. London: Chapman and Hall.

Treisman, J., Harris, E., Wilson, D. and Desplan, C. (1992). The homeodomain: a new face for the helix–turn–helix. *Bioessays* **14**, 145–150.

Vinson, C. R., Sigler, P. B. and McKnight, S. L. (1989). Scissors-grip model for DNA recognition by a family of leucine zipper proteins. *Science* **246**, 911–916.

Vrana, K. E., Churchill, M. A. E., Tullius, T. D. and Brown, D. D. (1988). Mapping functional regions of transcription factor TF III A. *Molecular and Cellular Biology* **8**, 1684–1696.

Walker, S., Hayes, S. and O'Hare, P. (1994). Site-specific conformational alteration of the Oct-1 POU domain–DNA complex as the basis for differential recognition by Vmw65 (VP16). *Cell* **79**, 841–852.

CHAPTER NINE

Activation and repression of gene expression by transcription factors

9.1 INTRODUCTION

Although binding to DNA is normally a necessary prerequisite for a factor to affect transcription, it is not in itself sufficient. Thus, after binding to DNA, the bound factor must interact with other transcription factors or with the RNA polymerase itself in order to affect the rate of transcription. Although such an interaction often results in the activation of transcription, this does not always occur, and a number of cases have now been defined in which the binding of a factor can result in repression of transcription. Activation and repression of transcription will, therefore, be discussed in turn.

9.2. ACTIVATION OF TRANSCRIPTION

9.2.1 Activation domains

Extensive studies on a variety of transcription factors have shown that they have a modular structure in which distinct regions of the protein mediate particular functions such as DNA binding (see Chapter 8) or interaction with specific effector molecules such as steroid hormones (see Section 4.3.1). It is likely, therefore, that a specific region of each individual transcription factor will be involved in its ability to activate transcription following DNA binding. As described in Section 2.3.3, such activation domains have been identified by so called 'domain-swap' experiments in which various regions of one factor are linked to the DNA-binding domain of another factor and the ability to activate transcription assessed.

In general, these experiments have confirmed the modular nature of transcription factors with distinct domains mediating DNA binding and transcriptional activation. Thus, in the case of the yeast factor GCN4, two distinct regions, each of 60 amino acids, have been identified which mediate, respectively, DNA binding and transcriptional activation (Figure 9.1a; Hope and Struhl, 1986). Similarly, domain-swap experiments have identified two regions of the glucocorticoid receptor, one at the N-terminus of the molecule and the other near the C-terminus, which can independently mediate gene activation when linked to the DNA-binding domain of another transcription factor (Hollenberg and Evans, 1988), and both of these are distinct from the DNA-binding domain of the molecule. Interestingly, the C-terminal activation domain is located close to the hormone-binding domain of the receptor (Figure 9.1b), and can mediate the activation of transcription only after hormone addition. It therefore plays an important role in the steroid-dependent activation of transcription following hormone addition (see Section 10.3.3).

Studies on a variety of transcription factors have therefore strongly

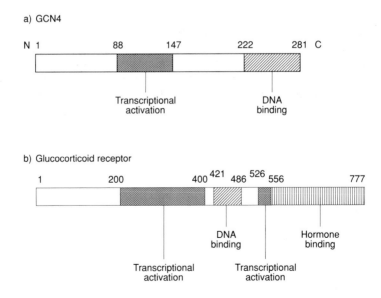

Figure 9.1 Domain structure of the yeast GCN4 transcription factor (a) and the mammalian glucocorticoid receptor (b). Note the distinct domains that are active in DNA binding or transcriptional activation.

indicated their modular nature, with distinct regions of the molecule mediating DNA binding and transcriptional activation. An extreme example of this modularity is provided by the interaction of the cellular transcription factor Oct-1 (see Sections 8.2 and 8.6) and the herpes simplex virus trans-activating protein VP16 (review: Goding and O'Hare, 1989). Thus, although VP16 contains a very strong activating region which can strongly induce transcription when artifically fused to the DNA-binding domain of the yeast GAL4 transcription factor, it contains no DNA-binding domain and cannot, therefore, bind to DNA itself. Transcriptional activation by VP16 following viral infection is therefore dependent upon its ability to form a protein–protein complex with the cellular Oct-1, protein. This complex then binds to the octamer-related TAATGARAT (R = purine) motif in the viral immediate-early genes via the DNA binding domain of Oct-1 and transcription is activated by the activation domain of VP16. Hence, in this case, the DNA-binding and transcriptional activation domains are actually located on different proteins in the DNA-binding complex (Figure 9.2). A similar example, in which the constitutively expressed Oct-1 recruits a non-DNA-binding, B cell-specific, activating molecule to the immunoglobulin promoter resulting in its B cell-specific activation, was discussed in Section 5.2.2 indicating that this effect is not confined to viral trans-activating molecules.

9.2.2. Nature of activation domains

Following the identification of activation domains in different transcription factors, it rapidly became clear that they fell into several distinct families with common features which will be discussed in turn (for a typical example of each of the major classes of activation domain see Figure 9.3).

Figure 9.2 Activation of gene transcription by interaction of the cellular factor Oct-1, which contains a DNA-binding domain, and the herpes simplex virus VP16 protein, which contains an activation domain but cannot bind to DNA.

GAL4 acidic domain

D S A A A H H D N S T I P L D F M P R D

A L H G F D W S E E D D M S D G L P F L

K T D P N N N G F

Sp1 glutamine-rich domain B

Q G Q T P Q R V S G L Q G S D A L N I Q

Q N Q T S G G S L Q A G Q Q K E G E Q N

Q Q T Q Q Q Q I L I Q P Q L V Q G G Q A

L Q A L Q A A P L S G Q T F T T Q A I S

Q E T L Q N L Q L Q A V P N S G P I I I

R T P T V G P N G Q V S W Q T L Q L Q N

L Q V Q N P Q A Q T I T L A P M Q G V S

L G Q

CTF/NFI Proline-rich domain

P P H L N P Q D P L K D L V S L A C D P

A S Q Q P G R L N G S G Q L K M P S H C

L S A Q M L A P P P P G L P R L A L P P

A T K P A T T S E G G A T S P S Y S P P

D T S P

Figure 9.3 Structure of typical members of each of the three classes of activation domains. Acidic, glutamine or proline residues are highlighted in the appropriate case.

Acidic domains

Comparison of several different activation domains, including those of the yeast factors GCN4 and GAL4 as well as the activation domain at the N-terminus of the glucocorticoid receptor and that of VP16, which were discussed above (Section 9.2.1), indicated that, although they do not show any strong amino acid sequence homology to each other, they all have a large proportion of acidic amino acids, producing a strong net negative charge (Figure 9.3). Thus the 82-amino-acid activating region of the glucocorticoid receptor contains 17 acidic residues (Hollenberg and Evans, 1988), whilst the same number of negatively charged amino acids is found within the 60-amino-acid activating region of GCN4 (Hope and Struhl, 1986). These findings indicated, therefore, that these activation regions consist of so-called 'acid blobs' or 'negative noodles' with a high proportion of negatively charged amino acids which are involved in the activation of transcription (reviews: Ptashne, 1988; Sigler, 1988; Hahn, 1993a).

In agreement with this idea, mutations in the activation domain of GAL4, which increase its net negative charge, increase its ability to activate transcription (Gill and Ptashne, 1987). Similarly, if recombination is used to create a GAL4 protein with several more negative charges, the effect on gene activation is additive, a mutant with four more negative charges than the parental wild type activating transcription nine fold more efficiently than the wild type (Gill and Ptashne, 1988).

Although the acidic nature of the activation domain is clearly important for its function, it is not the only feature required, since it is possible to decrease the activity of the GAL4 activation domain without reducing the number of negatively charged residues (Gill and Ptashne, 1987). Thus, other features of these domains, such as the presence of conserved hydrophobic residues or the ability to form a β-sheet structure, may be as important or even more important than the net negative charge (Hahn, 1993a; Leuther *et al.*, 1993).

It is clear, therefore, that, whilst a number of activation domains consist of an acidic region, the precise features of the domains required for activation are still uncertain. Although activation domains of this type form the majority of the activation domains so far identified in eukaryotic transcription factors from yeast to mammals, other types of activation domains have been identified in a number of different transcription factors in higher eukaryotes and these will be discussed in turn.

Glutamine-rich domains

Analysis of the constitutive transcription factor Sp1 (see Section 3.3.2) revealed that the two most potent activation domains contained approximately 25% glutamine residues and very few negatively charged residues (Courey and Tjian, 1988; Figure 9.3). These glutamine-rich motifs are essential for the activation of transcription mediated by these domains, since their deletion abolishes the ability to activate transcription. Most interestingly, however, transcriptional activation can be restored by substituting the glutamine-rich regions of Sp1 with a glutamine-rich region from the *Drosophila* homeobox transcription factor *Antennapedia* which has no obvious sequence homology to the Sp1 sequence. Hence, as with the acidic activation domains, the activating ability of a glutamine-rich domain is not defined by its primary sequence but rather by its overall nature in being glutamine rich. In agreement with this, a continuous run of glutamine residues with no other amino acids has been shown to act as a transcriptional activation domain (Gerber *et al.*, 1994).

Similar glutamine rich regions have been defined in transcription factors other than Sp1 and Antennapedia, including the N-terminal activation domains of the octamer-binding proteins Oct-1 and Oct-2, the *Drosophila* homeobox proteins Ultrabithorax and zeste, and the yeast HAP1 and HAP2 transcription factors, indicating that this motif is quite widespread, being found in different transcription factors in different species (review: Mitchell and Tjian, 1989).

Proline-rich domains

Studies on the constitutive factor CTF/NF1 which binds to the CCAAT box motif (see Section 3.3.3) defined a third type of activation domain distinct from those previously discussed. Thus the activation domain located at the C-terminus of CTF/NF1 is not rich in acidic or glutamine residues, but instead contains numerous proline residues, forming approximately one-quarter of the amino acids in this region (Mermod *et al.*, 1989; Figure 9.3). As with the other classes of activation domains, this region is capable of activating transcription when linked to the DNA-binding domains of other transcription factors. Moreover, as with the glutamine-rich domain, a continuous run of proline residues can mediate activation, indicating that the function of this type of domain depends primarily on its richness in proline (Gerber *et al.*, 1994). Similar proline-rich domains have now been identified in several other transcription factors such as the oncogene product Jun, AP2 and the C-terminal activation domain of Oct-2 (review: Mitchell and Tjian, 1989). Thus, as with the glutamine-rich domains, proline-rich domains are not confined to a single factor, whilst a single factor such as Oct-2 can contain two activation domains of different types.

In summary, therefore, it is clear that, as with DNA binding, several distinct protein motifs can activate transcription (Figure 9.3).

Functional relationship of the different activation domains

The existence of at least three distinct classes of activation domain raises the question of whether these three domains are functionally equivalent or whether they differ in their ability to activate transcription. This question was investigated by Seipel *et al.* (1992), who linked each of the activation domains to the DNA-binding domain of the GAL4 factor and tested the ability of these chimaeric proteins to activate transcription in mammalian cells when the GAL4 DNA-binding site was placed at different positions relative to the start site of transcription (Figure 9.4). In these experiments all three domains were able to

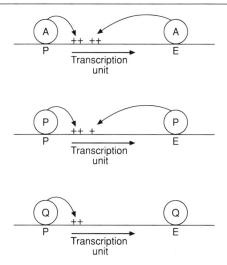

Figure 9.4 An acidic activation domain (A) can stimulate transcription when bound to DNA in the promoter (P) close to the transcriptional start site or when bound at a distant enhancer (E). In contrast, a proline-rich domain (P) stimulates only weakly from an enhancer position and a glutamine-rich domian (Q) does not stimulate at all from this position.

activate transcription when the DNA-binding site was placed close to the start-site of transcription in the promoter region. In contrast, the glutamine-rich domain was unable to activate transcription when the binding site was located downstream of the transcription unit, mimicking a position within an enhancer element (see Section 1.2.4). The acidic domain was strongly active from this enhancer position, whilst the proline-rich domain could also activate transcription from this position, but only weakly.

These findings indicate, therefore, that clear differences exist in the abilities of the different activation domains to activate transcription when bound to the DNA at different positions relative to the promoter. Such differences are likely to be important in determining the functional activity of different factors underlying, for example, the different effects of Oct-1 and Oct-2 following binding to the octamer motifs in the immunoglobulin promoter and enhancer, which were discussed in Section 5.2.2. In addition, such differences in the activity of different activation domains are likely to reflect differences in the mechanisms by which these factors act. In agreement with this idea, acidic or proline-rich activation domains derived from mammalian factors can also activate transcription when introduced into yeast cells, whereas glutamine-rich domains cannot do so (Kunzler *et al.*, 1994).

In the next sections, we will consider the mechanisms by which activation domains act, focusing particularly on the acidic domains where most information is available. Similarities and differences in the mode of action of the other activation domains will be discussed where this information is available (reviews: Horey and Carey, 1994; Tjian and Maniatis, 1994).

9.2.3 Interaction of activation domains with the basal transcriptional complex

The widespread interchangeability of acidic activation domains from yeast, *Drosophila* and mammalian transcription factors discussed above strongly suggests that a single common mechanism may mediate transcriptional activation by acidic activation domains in a wide range of organisms. This idea is supported by the finding that mammalian transcription factors carrying such domains, such as the glucocorticoid receptor, can activate a gene carrying their appropriate DNA-binding site in yeast cells, whilst the yeast GAL4 factor can do so in cells of *Drosophila*, tobacco plants and mammals (review Guarente, 1988; Ptashne, 1988).

Most interestingly, yeast and mammalian factors can co-operate together in gene activation, the rat glucocorticoid receptor and the yeast activator GAL4, for example, synergistically activating the transcription in mammalian cells of a gene containing binding sites for both factors (Kakidani and Ptashne, 1988). Such a synergistic interaction, in which activation by the yeast and mammalian factors together is far greater than the sum of that produced by either alone, is unlikely to be due to interactions between the two factors, which would normally never come into contact. Rather, it is likely to reflect their ability simultaneously to contact and activate another molecule (or molecules), which is part of the general transcription machinery common to yeast and mammalian cells. In agreement with this idea, synergistic activation by GAL4 and the mammalian factor ATF (Lin *et al.*, 1990) or by two bound molecules of GAL4 (Carey *et al.*, 1990) can be observed under conditions where binding sites on the DNA for these factors are saturated (Figure 9.5a). Hence such synergistic activation cannot be dependent upon co-operative binding of the two factors to DNA and must involve the ability of two different transcriptional activators or two molecules of the same factor to interact simultaneously with the same target factor (Figure 9.5b, c).

Such considerations clearly focus attention on the nature of the target factor or factors with which these activators interact. A number of experiments have indicated that this target factor is likely to be a component of the basal transcription complex (see Section 3.2) required for the transcription of a number of different genes and not solely for

a)

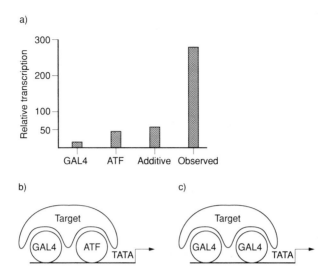

Figure 9.5 (a) Activation of a target promoter containing binding sites for the yeast transcription factor GAL4 and the mammalian transcription factor ATF. Note that the observed activation in the presence of both factors is greater than the additive effect of each factor individually, indicating that the two factors can activate gene expression synergistically. (b) and (c) Proposed mechanism for synergistic activation by GAL4 and ATF or by two molecules of GAL4 in which each molecule makes simultaneous contact with the target for activation.

that of the activated gene. Thus the over-expression of the yeast GAL4 protein, which contains a strong activation domain, results in the down-regulation of genes that lack GAL4-binding sites, such as the *CYC1* gene, as well as activating genes that do contain GAL4-binding sites (Gill and Ptashne, 1988). This phenomenon, which has been noted for a number of transcription factors with strong activation domains is known as squelching (review: Ptashne, 1988). Although the degree of squelching by any given factor is proportional to the strength of its activation domain, squelching differs from activation in that it does not require DNA binding and can be achieved with truncated factors containing only the activation domain and lacking the DNA-binding domain. This phenomenon can, therefore, be explained on the basis that a transcriptional activator, when present in high concentration, can interact with its target factor in solution as well as on the DNA. If this target factor is present at limiting concentrations, it will, therefore, be sequestered away from other genes that require it for transcription, resulting in their inhibition (Figure 9.6).

The existence of squelching indicates, therefore, that the target factor for activation domains is likely to be a component of the basal transcriptional apparatus which is required for the transcription of a wide range

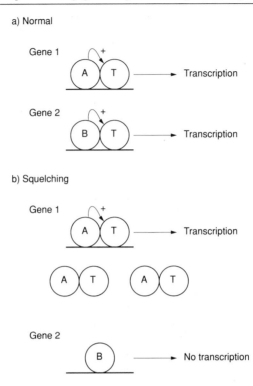

Figure 9.6 The process of squelching. In the normal case, illustrated in (a), two distinct activator molecules, A and B, involved in the activation of genes one and two respectively, both act by interacting with the general transcription factor T and both genes are transcribed. In squelching, illustrated in (b), factor A is present at high concentration and hence interacts with T both on gene 1 and in solution. Hence factor T is not available for transcription of gene 2 and therefore only gene 1 is transcribed whilst transcription of gene 2 is squelched.

of genes and which is conserved from yeast to mammals, allowing yeast activators to work in mammalian cells and vice versa. Obviously, an activating factor could act by stimulating the binding of such a component so that the basal complex assembled more efficiently. Alternatively, it could act by interacting with a factor that had already bound so that the activity or stability of the assembled complex was stimulated. It appears that both these mechanisms are used and they will be discussed in turn.

Factor binding

As described in Section 3.2.4, the binding of TFIID to the promoter is followed by binding of TFIIB, which in turn recruits the RNA polymer-

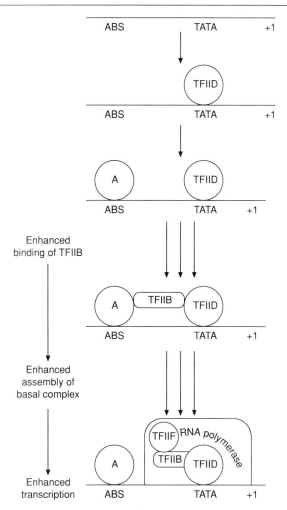

Figure 9.7 Following binding of TFIID to the TATA box, the binding of an activating molecule (A) to its binding site (ABS) will enhance the recruitment of the TFIIB factor, so enhancing the rate of basal complex assembly and of transcription.

ase and associated factors. It has been shown that both an acidic activator and glutamine- or proline-rich activators can greatly stimulate the binding of TFIIB to the promoter (Choy and Green, 1993). Hence, once TFIID has bound, the subsequent binding of an activator to its binding site in the promoter will greatly stimulate the rate at which TFIIB binds, so promoting assembly of the basal transcription complex (Figure 9.7).

Although the increased recruitment of TFIIB will evidently result in increased assembly of the complex, simply because more TFIIB is bound

LIVERPOOL JOHN MOORES UNIVERSITY
LEARNING SERVICES

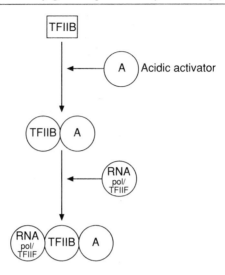

Figure 9.8 The binding of an acidic activator (A) to TFIIB produces a conformational change which enhances the ability of TFIIB to interact with the RNA polymerase–TFIIF complex, thereby enhancing its recruitment to the promoter.

to the promoter, there is evidence that activators also directly increase the ability of bound TFIIB to recruit the complex of RNA polymerase and TFIIF to the promoter (Roberts and Green, 1994). Thus it has been shown that the interaction of TFIIB, with an acidic activator results in a conformational change in TFIIB which increases its ability to bind the RNA polymerase–TFIIF complex (Figure 9.8).

Hence the interaction of activators with TFIIB both enhances its rate of recruitment to the promoter, as well as directly improving its ability to recruit the other factors that bind subsequently. These alterations in TFIIB have been shown to depend on the direct interaction of TFIIB with activating molecules. Thus TFIIB can be purified on a column containing a bound acidic activator, and interactions of TFIIB with non-acidic activators have also been reported. Moreover, mutations in the activator that abolish this interaction with TFIIB prevent it from activating transcription (review: Sharp, 1991; Hahn, 1993b). Thus the effect of activators on TFIIB is mediated via a direct protein–protein interaction which is essential for their ability to stimulate transcription.

Factor activity

Although the ability of activators to enhance the assembly of the basal transcriptional complex is clearly essential for transcriptional activa-

Figure 9.9 An activator can stimulate transcription both by promoting the assembly of the basal transcription complex and by stimulating its activity following assembly.

tion, it is not the only mechanism by which activators can stimulate transcription. Thus activators can also stimulate transcription at a subsequent step following assembly of the complex, resulting in its enhanced stability or increased activity (Choy and Green, 1993) (Figure 9.9). Whilst it is possible that this subsequent stage also involves interaction with TFIIB, it is more likely that it involves other factors within the complex.

An obvious mechanism for activation would be for activating domains to interact directly with the RNA polymerase itself to increase its activity (Figure 9.10a). As discussed in Section 3.1, the largest subunit of RNA polymerase II contains multiple copies of a sequence whose consensus is Tyr-Ser-Pro-Thr-Ser-Pro-Ser, which is highly conserved in evolution and is essential for its function. This motif is very rich in hydroxyl groups and lacks negatively charged acidic residues.

It has, therefore, been suggested (Sigler, 1988) that this motif could interact directly, either with a negatively charged acidic domain or via

Figure 9.10 Two possible mechanisms by which an activating factor (A) could stimulate the activity of the basal trascriptional complex. This could occur via direct interaction with the RNA polymerase itself (a) or by interaction with another transcription factor such as TFIID, which in turn interacts with the polymerase (b).

hydrogen bonding with amide groups in a glutamine-rich activation domain. This would provide a mechanism for direct interaction between activating domains and RNA polymerase itself, whilst the evolutionary conservation of the target region within the polymerase would explain why yeast activators work in mammalian cells and vice versa. In agreement with this idea it has been shown that yeast mutants containing a reduced number of copies of the heptapeptide repeat in RNA polymerase II are defective in their response to activators such as GAL4 (Scafe *et al.*, 1990).

Although these results are consistent with a direct interaction between transcriptional activators and the RNA polymerase, they are equally consistent with an indirect interaction in which the activator contacts another component of the transcriptional machinery which then interacts with the repeated motif in the polymerase. Indeed, it has recently been shown that the interaction of activators with RNA polymerase *in vitro* requires a mediator complex consisting of some 20 polypeptides rather than occurring directly (Kim *et al.*, 1994). Therefore, despite the attractiveness of a model involving direct interaction between activating factors and the polymerase itself, it is unlikely to be correct, and it appears that activators interact with the polymerase indirectly via another factor (Figure 9.10b). TFIID is an obvious candidate for the component with which activating factors interact, since this factor is required for the transcription of a wide variety of genes, both with and without TATA boxes (see Section 3.2.5), and is highly conserved in evolution, with the yeast factor being able to promote transcription in mammalian cell extracts and vice versa.

Evidence for an effect of activating factors on TFIID has been obtained in the case of the yeast acidic activating factor GAL4 (Horikoshi *et al.*, 1988a). Thus, in the absence of GAL4, TFIID was shown to be bound only at the TATA box of a promoter containing both a TATA box and GAL4-binding sites. In contrast, in the presence of GAL4 bound to its upstream binding sites in the promoter, the conformation of TFIID was altered such that it now covered both the TATA box and the start site for transcription (Figure 9.11). Moreover, no change in TFIID conformation was observed in the presence of a truncated GAL4 molecule which can bind to DNA but lacks the acidic activation domain. Hence, an acidic activator can produce a change in TFIID conformation, resulting in its binding to the start site for transcription, and this effect correlates with the ability of GAL4 to activate transcription rather than being a consequence of its binding to DNA. A similar change in the binding of TFIID has also been noted following the binding of the mammalian activating factor ATF to sites upstream of the TATA box (Horikoshi *et al.*, 1988b), suggesting that TFIID is the common target for GAL4 and ATF detected in the experiments of Lin *et al.* (1990) discussed above. The observed change in TFIID conformation in the presence of GAL4 or ATF would presumably result in a change in the basal transcriptional complex, resulting in its increased activity or stability.

It is clear, therefore, that activating molecules can alter the configuration of TFIID bound to the promoter by interacting directly with it. As with TFIIB, TBP (the DNA-binding component of TFIID; see Section

Figure 9.11 Effect of GAL4 binding on the binding of TFIID.

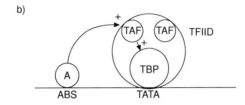

Figure 9.12 Interaction of an activator molecule with TBP can occur either directly (a) or indirectly (b) via an intermediate TBP-associated adaptor molecule (TAF).

3.2.5) can interact directly with acidic activators, and single amino acid mutations in such activators, which prevent interaction with TBP, also abolish the ability to stimulate transcription. Although there is thus evidence that the ability to interact with TBP appears to be essential for transcriptional activation (Figure 9.12a), there is also evidence that in some circumstances such activation requires interaction of the activator with the TAFs (TBP-associated factors) which, together with TBP, make up the TFIID complex (see Section 3.2.5).

Thus, in many cases, stimulation of transcription *in vivo* by activator molecules does not occur with purified TBP but is dependent upon the presence of the TFIID complex and hence of the TAFs. This suggests a model in which the interaction of activators with TBP occurs indirectly via TAFs, which have also, therefore, been called adaptor molecules (Figure 9.12b) (reviews: Goodrich and Tjian, 1994; Tjian and Maniatis, 1994). In agreement with this, the yeast Sug1 protein, which was identified on the basis that mutations in the gene encoding it affect activator function, has been shown to interact directly with both acidic activation domains and TBP (Swaffield *et al.*, 1995). Similarly, the mammalian homologue of this protein was recently cloned on the basis of its ability to interact with the transcription activation domain of the thyroid hormone receptor (Lee *et al.*, 1995). The critical role of such interactions is shown by the finding that a mutation which inactivates the TBP-associated factor TAF$_{II}$ 250 inhibits the expression of specific genes and results in cell cycle arrest (Wang and Tjian, 1994).

Interestingly, there is now evidence that different classes of activation

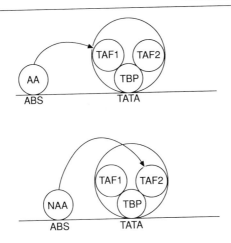

Figure 9.13 Acidic (AA) and non-acidic (NAA) activator molecules may interact with different TBP-associated factors (TAFS) within the TFIID complex.

domain may interact with different TAFs (Chen *et al.*, 1994). Thus, whilst an acidic activation domain has been shown to interact directly with $TAF_{II}40$, the glutamine-rich domain of Sp1 interacts with $TAF_{II}110$. Hence acidic and non-acidic activation domains may have different targets within the TFIID complex (Figure 9.13).

In agreement with this idea, the acidic activation domain of VP16 is not capable of squelching gene activation by the non acidic activation domain of the oestrogen receptor, whereas the oestrogen receptor activation domain is capable of squelching gene activation mediated both by its own activation domain and by the acidic domain of VP16, indicating that they contact different molecules (Tasset *et al.*, 1990). Moreover, these findings suggest that a series of TAFs within TFIID may mediate activation, with the acidic activation domain of VP16 contacting a factor that is located earlier in the series than that contacted by the non-acidic activation domain of the oestrogen receptor (Figure 9.14). Hence the factor contacted by the activation domain of the oestrogen receptor would also be essential for activation by VP16 (factor 4 in Figure 9.14), whereas the factor contacted by the acidic activation domain of VP16 (factor 1 in Figure 9.14) would not be required for activation by the oestrogen receptor.

The functional differences that exist between different factors in their ability to activate transcription from different positions and in different species (see Section 9.2.2) are, therefore, paralleled by differences in their ability to interact with different TAFs.

There thus exists an array of target factors that are contacted by transcriptional activators; these include TFIIB, TBP and the various

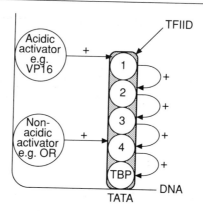

Figure 9.14 Interaction of different activator molecules with different adaptor molecules (1–4) which each activate each other and ultimately activate TBP. Note that the ability of the non-acidic activation domain of the oestrogen receptor to squelch activation by the acidic activation domain of VP16, but not vice versa, can be explained if the oestrogen receptor interacts with an adaptor molecule (4) closer to TBP in the series than that with which VP16 interacts (1).

TAFS, with some factors being contacted by activators of all classes and others by activators of only one class. Indeed, this array has recently been extended by the findings that TFIIA (Ozer *et al.*, 1994), TFIIF (Joliot *et al.*, 1995) and TFIIH (Xiao *et al.*, 1994) can interact with transcriptional activators. Even when the finding that some of these targets, such as individual TAFs, can interact with only one class of activation domain is taken into account, there still remains a bewildering number of targets within the basal complex. It should be noted, however, that the various possible targets are not mutually exclusive. Indeed, the ability of different molecules of the same factor or different activating factors to interact with different components within the basal transcriptional complex is likely to be essential for the strong enhancement of transcription, which is the fundamental aim of activating molecules (Figure 9.15).

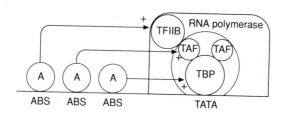

Figure 9.15 The ability of multiple activating molecules (A) to contact different components of the basal transcriptional complex allows strong activation of transcription.

9.2.4 Other targets for transcriptional activators

Although the basal transcriptional complex that initiates transcription is the best characterized target for transcriptional activators (as discussed in the preceding section), at least two other stages of the transcriptional process can be targeted by such activators and these will be discussed in turn.

Modulation of chromatin structure

In the cell, the DNA molecule is associated with histones and other proteins to form particles known as nucleosomes, which are the basic unit of chromatin structure. Prior to the onset of transcription, the chromatin structure becomes altered, thus allowing the subsequent binding of the factors that actually stimulate transcription (reviews: Felsenfeld, 1992; Latchman, 1995). As discussed in previous chapters, this alteration in chromatin structure can itself be produced by the binding of a specific transcription factor. This results in a change in the nucleosome pattern of DNA–histone association, thereby allowing

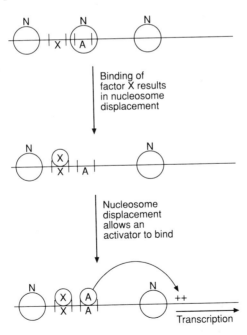

Figure 9.16 A transcription factor (X) can stimulate transcription by binding to DNA and displacing a nucleosome (N), so allowing a constitutively expressed activator (A) to bind and activate transcription.

other activating factors access to their specific DNA-binding sites (Figure 9.16) (reviews: Croston and Kadonaga 1993; Lewin 1994; Paranjape *et al.*, 1994; Travers, 1994).

Thus, as described in Section 4.3.3, the binding of the activated glucocorticoid receptor to the mouse mammary tumour virus promoter results in an alteration in the nucleosomal organization of the DNA. This allows the ubiquitous NF1 factor to bind to its binding site, which was previously masked by a nucleosome, and hence allows it to activate transcription by interacting with the basal transcriptional complex. Similarly, as discussed in Section 4.2.1, the binding site for the heat shock factor (HSF) in the hsp70 gene promoter is unmasked by the prior binding of the GAGA factor to its binding site in the gene promoter. In non-heat-shock cells, however, HSF is present in an inactive non-DNA-

Figure 9.17 Two-stage induction of genes by steroid treatment or heat shock, involving alteration of chromatin structure and the subsequent binding of an activator molecule. In the steroid case (a), the steroid activates the glucocorticoid receptor (GR), resulting in a change in chromatin structure and immediate binding of the constitutively expressed NF1 factor. In contrast, in the heat-inducible case (b), the chromatin structure has already been altered prior to heat shock by binding of the GAGA factor but transcription occurs only when heat shock activates the heat-shock factor (HSF) to a DNA-binding form.

binding form and therefore cannot bind to its site even though it is accessible. Following heat shock, HSF is converted to a DNA-binding form and immediately binds to its site and activates transcription.

These examples illustrate, therefore, how factors that alter chromatin structure can prepare the way for the binding of the factors that actually activate transcription, with such binding occurring either immediately following the change in chromatin structure, as in the glucocorticoid receptor–NF1 case, or following a subsequent stimulus, as in the GAGA–HSF case (Figure 9.17). The importance of factors that can alter chromatin structure is illustrated by the recent finding that the mutational inactivation of the gene encoding the GAGA factor in *Drosophila* leads to reduced viability, eye abnormalities and female sterility (Farkas *et al.*, 1994). This presumably results from a failure to express specific genes which are critical for development and whose activation is normally dependent on the reprogramming of their chromatin structure by the GAGA factor.

Interestingly, as well as interacting with the basal transcriptional apparatus, activation domains have also been shown to be involved in the ability of specific factors to alter chromatin structure. Thus, the activation of the yeast PHO5 gene promoter following phosphate starvation is mediated by the binding of the PHO4 factor to the PHO5 promoter, resulting in nucleosome displacement. Surprisingly, a truncated PHO4 molecule, lacking the acidic activation domain but retaining the DNA-binding domain, is incapable of nucleosome displacement, whilst this ability can be restored by linking the truncated PHO4 molecule to the acidic activation domain of VP16 (Figure 9.18). Hence the ability of PHO4 to disrupt chromatin structure is dependent upon the acidic activation domain, which also interacts with the basal transcription complex to stimulate transcription (Suaren *et al.*, 1994). This

Figure 9.18 Both the disruption of chromatin and the activation of the basal transcription complex by the yeast PHO4 factor are dependent on its acidic activation domain. They are therefore lost when this domain is deleted but can be restored by addition of the acidic activation domain of VP16.

dual function is also seen in the case of the glucocorticoid receptor which, as well as altering chromatin structure thereby facilitating NF1 binding and consequent transcriptional activation, can also itself directly stimulate transcription via interaction with other transcription factors (Section 4.3.3).

These findings indicate, therefore, that, at least in the case of the acidic domains, an activation domain can interact with chromatin components to modulate chromatin structure as well as interacting with the basal transcriptional apparatus to stimulate transcription.

Stimulation of transcriptional elongation

In most genes, once transcription has been initiated the RNA polymerase continues to transcribe the DNA until it has produced a complete RNA transcript. In some genes, however, such as the c-*myc* oncogene (see Section 7.2.4), some transcripts terminate prematurely and do not produce an RNA capable of encoding the appropriate protein (review: Greenblatt *et al.*, 1993). Moreover, this process can be regulated. Thus when the c-*myc* gene is transcribed in the pro-myeloid cell line HL-60 most transcripts terminate near the end of the first exon and do not produce a functional mRNA encoding the complete Myc protein. When the HL60 cells are differentiated to form granulocytes, however, the majority of trancripts pass through this block and full-length mRNA is produced (Figure 9.19). Hence, in this case, an increased level of functional c-*myc* mRNA able to produce the Myc protein is obtained without an increase in transcriptional initiation (review: Spencer and Groudine, 1990).

Although the regulatory factor that modulates c-*myc* elongation is unknown, a factor acting at the level of transcriptional elongation has been identified in the human immunodeficiency virus (HIV). Thus, in this case, only short, prematurely truncated, transcripts are produced from the HIV promoter in the absence of the viral Tat protein. When Tat is present, however, it stimulates both the rate of initiation of transcription and also the proportion of full-length transcripts produced, so overcoming the block to transcriptional elongation (Figure 9.20) (review: Greenblatt *et al.*, 1993). A similar role for HSF in stimulating transcriptional elongation in the *hsp70* gene has also been proposed (review: Lis and Wu, 1993). Hence activating factors can act to stimulate the proportion of full-length RNA transcripts that are produced.

Interestingly, as in the case of transcriptional initiation and the disruption of chromatin, there is evidence that the stimulation of transcriptional elongation involves activation domains. Thus, when binding sites for the yeast transcriptional activator GAL4 were placed upstream

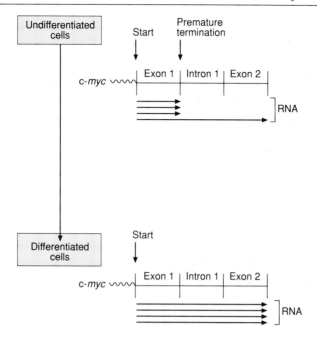

Figure 9.19 In undifferentiated HL60 cells, most transcripts from the c-*myc* gene terminate prematurely at the end of the first exon. In differentiated cells, however, this does not occur, resulting in an increase in full-length functional transcripts without increased initiation.

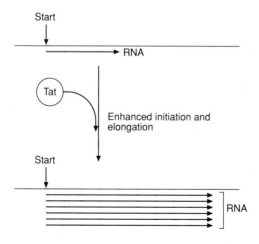

Figure 9.20 The human immunodeficency virus Tat protein stimulates transcriptional initiation so that more RNA is made, and also enhances the proportion of full-length RNA species capable of encoding viral proteins.

of the c-*myc* promoter, both the rate of initiation and the proportion of full-length transcripts were greatly stimulated by the binding of hybrid transcription factors containing the DNA-binding domain of GAL4 linked to an acidic or non-acidic activation domain (Yankulov *et al.*, 1994). This effect was not observed in the presence of the GAL4 DNA-binding domain alone or when the GAL4 binding sites were deleted (Figure 9.21). Hence the ability of activating factors to act at the level of transcriptional elongation is dependent on the same activation domains that act to stimulate transcription at other stages.

These findings indicate, therefore, that, in the minority of genes where some transcripts terminate prematurely, it is possible for transcriptional activators to act by enhancing the proportion of full-length transcripts that are produced, as well as by stimulating the number of transcripts initiated.

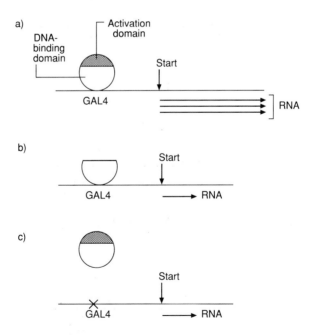

Figure 9.21 (a) The binding of a chimaeric transcription factor containing the GAL4 DNA-binding domain linked to an activation domain (shaded) stimulates both transcriptional initiation and the proportion of full-length transcripts produced by a c-*myc* gene carrying a DNA-binding site for GAL4. This effect is not observed when only the DNA-binding domain of GAL4 is present (b) or when the GAL4-binding sites are deleted so that the activator cannot bind (c).

9.2.5 Multi-stage activation of transcription

It is clear, therefore, that specific activation domains within transcription factors can act at three distinct stages to stimulate transcription. Thus they can disrupt the chromatin structure to allow other activating factors to bind, stimulate the rate of transcriptional initiation so that more RNA transcripts are initiated and, in some genes, can increase the proportion of full-length transcripts produced (Figure 9.22). Taken

Figure 9.22 An activation domain (hatched) within an activating transcription factor can act (a) by disrupting the nucleosome (N) arrangement of the DNA so that an activating factor can bind; (b) by stimulating the rate of transcriptional initiation; and (c) by increasing the proportion of initiated transcripts that go on to produce a full-length RNA.

together with the multiple targets for activating factors within the initiation complex, these effects allow transcriptional activators to fulfil their function and strongly stimulate transcription.

9.3 REPRESSION OF TRANSCRIPTION

Although the majority of transcription factors which have so far been described act in a positive manner, a number of cases have now been reported in which a transcription factor exerts an inhibitory effect on transcription. This effect can occur by indirect repression, in which the repressor interferes with the action of an activating factor, so preventing it stimulating transcription (Figure 9.23a–c). Alternatively, it can occur via direct repression, in which the factor reduces the activity of the basal transcriptional complex (Figure 9.23d). These two mechanisms will be discussed in turn (reviews: Goodbourn, 1990; Clark and Docherty, 1993; Herschbach and Johnson, 1993).

9.3.1 Indirect repression

Several mechanisms exist by which an inhibitor can interfere with the action of an activator.

Inhibition of activator binding by masking of its DNA-binding site

One means by which repression can occur is by the masking of the DNA-binding site for the factor, so preventing it binding to the DNA and activating transcription. By preventing the binding of the positively acting factor, the negatively acting factor effectively inhibits gene activation. This masking of the binding site can be achieved simply by the negatively acting factor binding to the same site as the positively acting factor but failing to activate transcription (Figure 9.24a). This is seen, for example, in the case of the Sp3 factor, a newly discovered factor related to the Sp1 factor discussed in Section 3.3.2. Thus the Sp3 factor binds to the same site as Sp1 but, unlike Sp1, it cannot activate transcription. It therefore blocks the Sp1-binding site, preventing Sp1 binding and activation of transcription (Hagen *et al.*, 1994).

Other examples of this mechanism have been discussed in previous chapters. Thus, the Eng protein binds to the same site as the Ftz protein, and hence Ftz cannot stimulate gene expression in the presence of Eng (see Section 6.2.3). Similarly, the binding of the gluco-

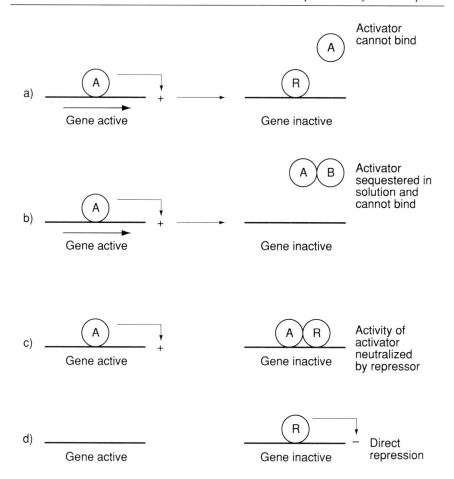

Figure 9.23 Potential mechanisms by which a transcription factor can repress gene expression. This can occur (a) by the repressor (R) binding to DNA and preventing an activator (A) from binding and activating gene expression; (b) by the repressor interacting with the activator in solution and preventing its DNA binding; (c) by the repressor binding to DNA with the activator and neutralizing its ability to activate gene expression; or (d) by direct repression by an inhibitory transcription factor.

corticoid receptor to an nGRE inhibits gene expression by preventing activator binding (see Section 4.3.4), whilst the CCAAT displacement protein prevents the CCAAT-binding protein from binding to its site in the sperm histone H2B gene in embryonic tissues (see Section 3.3.3).

As well as directly masking the DNA-binding site of an activator by binding directly to the site, an inhibitor can also achieve such masking indirectly by reorganizing the nucleosome arrangement of the chromatin. This effect is evidently the reverse of the mechanism of gene activation, which involves nucleosome displacement, so unmasking an

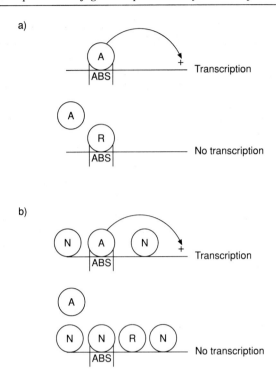

Figure 9.24 A repressor (R) can prevent gene expression by masking the binding sites (ABS) for an activator (A), either by binding directly to the site (a) or by reorganizing chromatin structure so that the binding site is masked by a nucleosome (N) (b).

activator-binding site and allowing it to bind (see Section 9.2.4). Thus, if a factor reorganizes the pattern of nucleosomes so that the binding site of the activator is masked by a nucleosome, this will effectively repress gene activation (Figure 9.24b). An example of a factor that acts in this way is the polycomb repressor of *Drosophila*, which normally represses inappropriate expression of several homeotic genes by modulating their chromatin structure so that activating molecules cannot bind (Franke *et al.*, 1992). When this factor is inactive, inappropriate expression of these genes in the wrong cell type is observed, leading to dramatic transformations in the nature of specific parts of the body.

Inhibition of activator binding by formation of a non-DNA-binding complex

In addition to repressing gene expression by binding to an nGRE, the glucocorticoid receptor also illustrates another mechanism of gene repression, namely the formation of a non-DNA-binding complex with an activating factor (Figure 9.23b). Thus, as discussed in Section 4.3.4, the glucocorticoid receptor can interact with the AP1 factor to form a non-DNA-binding complex, thereby preventing AP1 from binding to DNA and activating transcription. Obviously, such an interaction will also prevent the glucocorticoid receptor from binding to the DNA of genes that it can activate resulting in mutual trans-repression of both glucocorticoid and AP1-dependent genes. Hence the glucocorticoid receptor can both activate the expression of some genes (see Section 4.3.1) and repress the expression of other genes, either by binding to DNA in a configuration that cannot activate transcription or by forming a non-DNA-binding complex with another activating factor.

As well as cases of this type, in which a single factor can act either positively or negatively depending on the gene involved, gene repression involving the inhibition of DNA binding by protein–protein interaction can also be produced by factors that function purely negatively. Thus, as discussed in Section 5.3.2, gene activation by helix–loop–helix proteins can be inhibited by the Id protein, which can dimerize with them by means of its helix–loop–helix domain and then inhibit their DNA-binding since it lacks the basic DNA-binding domain.

Quenching of an activator

The cases of repression described so far all involve the inhibition of DNA binding either by blocking the binding site for a factor (Figure 9.23a) or by forming a non-DNA-binding protein–protein complex (Figure 9.23b). Since DNA binding is a necessary prerequisite for gene activation, this constitutes an effective form of repression. In addition, however, inhibition of transcription can also be achieved by interfering with transcriptional activation by a DNA-bound factor in a phenomenon known as quenching (Figure 9.23c). A simple example of this type is seen in the case of the negatively acting yeast factor GAL80, which inhibits gene activation by the positively acting GAL4 protein. This is achieved by the binding of GAL80 to DNA-bound GAL4, such binding occurring via the 30 amino acids located at the extreme C-terminus of the GAL4 molecule. As these amino acids are located close to the GAL4 activation domain, the binding of GAL80 to GAL4 masks the GAL4 activation domain and hence inhibits the activation of gene expression by GAL4. In response to treatment with galactose, GAL80 dissociates from GAL4, allowing GAL4 to fulfil its function of activating galactose-inducible genes. Hence this system provides an elegant means of modulating gene expression in response to galactose, with the activating GAL4 factor being bound to DNA both prior to and after galactose addition, but

Figure 9.25 Galactose activates gene expression by removing the GAL80 protein from DNA-bound GAL4 protein, unmasking the activation domain on GAL4.

being able to activate gene expression only following the galactose-induced dissociation of the quenching GAL80 factor (Johnston *et al.*, 1987) (Figure 9.25). A similar case, in which the activation domain of p53 is masked by the binding of the MDM2 protein to p53, was discussed in Section 7.3.2.

A related example of quenching, in which the inhibitory factor binds to a DNA sequence adjacent to the quenched factor rather than only to the factor itself, is seen in the case of the c-*myc* promoter. Thus an inhibitory transcription factor, myc-PRF, binds to a site adjacent to that occupied by an activating factor, myc-CF1, and interferes with its ability to activate c-*myc* gene transcription (Kakkis *et al.*, 1989). Hence quenching can occur either by an inhibitory factor binding to the positively acting factor (Figure 9.26a) or by the inhibitory factor binding to DNA adjacent to the positive factor (Figure 9.26b). In both cases, however, this effect involves the inhibitor interfering with the ability of the activator's activation domain to stimulate transcription.

9.3.2 Direct repression

In the cases described so far, a negative factor exerts its effect by neutralizing the action of a positively acting factor by preventing either its DNA binding (Figure 9.23a, b) or inhibiting its activation of transcription following such binding (Figure 9.23c). In other cases, however, the inhibitory effect of a particular factor can be observed in the absence

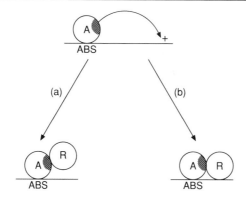

Figure 9.26 The ability of a bound activator (A) to stimulate transcription via its activation domain (hatched) can be inhibited by quenching of the activation domain by inhibitory factors (R) which either bind to the activator without binding to DNA (a) or which bind to a site adjacent to the activator (b).

of any activating factors. This indicates that these inhibitory factors inhibit transcription directly by interacting with the basal transcriptional complex to reduce its activity (Figure 9.23d). Several factors of this type have now been shown to bind to specific DNA-binding sites within their target genes and reduce the activity of the basal transcriptional complex (Figure 9.27a). This effect is evidently similar in nature but opposite in effect to the stimulation of the basal complex by the binding of activating factors to specific DNA sequences in the promoter.

A number of factors capable of inhibiting the basal initiation complex

Figure 9.27 An inhibitory factor (R) can reduce the activity of the basal transcriptional complex either by binding to DNA and then interacting with the complex (a) or by binding directly to the complex by protein–protein interaction (b).

in this way have been described in previous chapters, including the yeast α2 factor (see Section 5.4.3), the *Drosophila* eve protein (see Section 6.2.2) and the mammalian thyroid hormone receptor, which can directly inhibit the basal complex in the absence of thyroid hormone but stimulates its activity in the presence of the hormone (see Section 4.3.4). Interestingly, inhibitory factors of this type generally contain a small domain that can confer the ability to repress gene expression upon the DNA-binding domain of another factor when the two are artifically linked (e.g. Baniahmad *et al.*, 1992; Han and Manley, 1993; Lillycrop *et al.*, 1994) (Figure 9.28).

Hence these directly repressing factors contain specific inhibitory domains, paralleling the existence of specific activation domains in activating transcription factors. By analogy with activation domains, inhibitory domains are likely to inhibit either the assembly of the basal transcriptional complex or reduce its activity and/or stability after it has assembled. In agreement with this idea, the inhibitory domain of the thyroid hormone receptor has been shown to interact directly with TFIIB (Baniahmad *et al.*, 1993).

Interestingly, the inhibitory domain in the human Wilms tumour anti-oncogene product, and those from several *Drosophilia* inhibitory factors including eve, appear to share the common features of proline richness and an absence of charged residues (Han and Manley, 1993), suggesting that these factors have a common inhibitory domain. However, other inhibitory domains, such as those in the mammalian factors Oct-2 (Lillycrop *et al.*, 1994) and E4BP4 (Cowell and Hurst, 1994), are

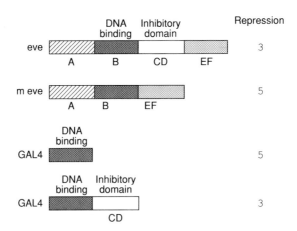

Figure 9.28 A specific region (CD) of the eve factor, which is distinct from the DNA-binding domain (B), acts as a transferable inhibitory domain. Thus its deletion from the eve protein results in a loss of the ability to repress transcription, whilst its linkage to the DNA-binding domain of GAL4 generates a functional repressor.

distinct both from this common domain and from each other, indicating that, as with activation domains, several types of inhibitory domain may exist.

Whatever the precise number of inhibitory domains, it is clear that a number of factors possess such domains and can, therefore, inhibit transcription by binding to upstream DNA sequences and inhibiting the activity of the basal transcriptional complex. The binding of such factors is likely to be of vital importance in producing the inhibitory effect of many of the silencer elements described in Section 1.3, the silencer element in the chicken lysozyme gene, for example, having been shown to act by binding the inhibitory thyroid hormone receptor.

As well as interfering with the basal complex by binding to distinct DNA-binding sites (Figure 9.27a), it is also possible for inhibitory factors to bind to the complex itself by protein–protein interaction and thereby interfere with its activity or assembly (Figure 9.27b) (review: Drapkin *et al.*, 1993). An example of this is provided by the Dr1 protein, which inhibits the assembly of the basal transcriptional complex by binding to TBP and preventing TFIIB from binding (Figure 9.29) (Inostroza *et al.*, 1992). As the recruitment of TFIIB to the promoter by interaction with the TBP component of TFIID is an essential step in

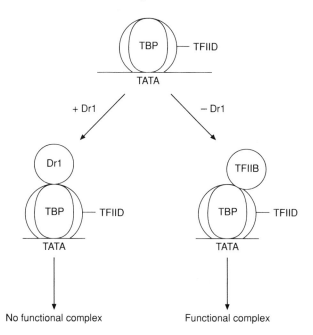

Figure 9.29 The Dr1 inhibitory factor can interact with the TBP component of TFIID, thereby preventing it from binding TFIIB and thus inhibiting the assembly of the basal transcriptional complex.

the assembly of the basal transcriptional complex, this effectively inhibits transcription (see Section 3.2.4).

As noted in Section 3.2.5, TBP is a component of the initiation complexes of all three polymerases, and in each case acts by recruiting other factors to the promoter. It has recently been shown (White *et al.*, 1994) that Dr1 can inhibit this ability of TBP to recruit other factors within the RNA polymerase II and III initiation complexes, but not within the RNA polymerase I complex. It therefore inhibits transcription by RNA polymerase II and III but not by RNA polymerase I. Thus Dr1 may play a critical role in regulating the balance of transcriptional activity between the ribosomal genes, which are the only genes transcribed by RNA polymerase I, and all the other genes in the cell.

In addition to Dr1, other factors that bind to TBP and inhibit the assembly of the RNA polymerase II basal complex, such as NC1, NC2 and Dr2, have been described and are likely to be important in controlling the rate of transcription (review: Drapkin *et al.*, 1993). Indeed, as well as interacting with activating factors (see Section 9.2.2), the TFIIA factor also appears to be able to bind to TBP, preventing these inhibitors from binding, thus allowing the recruitment of TFIIB (Figure 9.30) (see Section 3.2.4). This indicates that the activity of inhibitory molecules, which act by interacting with the basal transcriptional complex, can be regulated by activating factors and, once again, illustrates that the balance between transcriptional activators and repressors is of central importance in the control of transcription.

9.4 CONCLUSIONS

The fundamental property of transcription factors is their ability to influence the transcription of specific genes either positively or negatively. In this chapter, we have discussed how the activation of transcription is dependent on discrete activation domains whose structure and mechanism of action are gradually being understood, whilst the repression of transcription can be produced either by direct repression or by the neutralization of a positively acting factor via the inhibition of its DNA binding or activating properties. These properties offer ample scope for gene regulation in different cell types or in different tissues. Thus, in addition to the simple activation of gene expression by a positively acting factor present in only one cell type, the effect of a positively acting factor present in several different cell types can be affected by the presence or absence of a negatively acting factor which

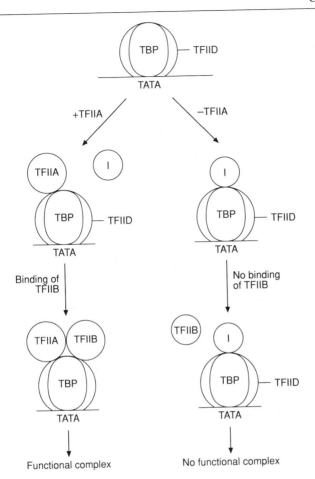

Figure 9.30 Binding of TFIIA to TBP prevents inhibitory molecules (I) from binding but still allows the binding of TFIIB and thereby promotes the assembly of the basal transcriptional complex.

is active in only one cell type and which inhibits its activity. Similarly, a single factor may act either positively or negatively, depending on the gene involved (as in the case of the glucocorticoid receptor) or depending on whether a specific hormone is present (as in the case of the thyroid hormone receptor).

Ultimately, however, all such potential mechanisms of gene regulation in response to specific stimuli or in specific cell types are dependent upon mechanisms that control the synthesis or activity of specific transcription factors in different cell types or in response to specific stimuli. These mechanisms are discussed in Chapter 10.

REFERENCES

Baniahmad, A., Kohne, A. C. and Renkawitz, R. (1992). A transferable silencing domain is present in the thyroid hormone receptor, in the v-*erbA* oncogene product and in the retinoic acid receptors. *EMBO Journal* **11**, 1015–1023.

Baniahmad, A., Ha, I., Reinberg, D., Tsai, S., Tsai, M.-J. and O'Malley, B.W. (1993). Interaction of human thyroid hormone receptor β with transcription factor TFIIB may mediate target gene derepression and activation by thyroid hormone. *Proceedings of the National Academy of Sciences USA* **90**, 8832–8836.

Carey, M., Lin, Y.-S., Green, M. R. and Ptashne, M. (1990). A mechanism for synergistic activation of a mammalian gene by GAL4 derivatives. *Nature* **345**, 361–364.

Chen, J.-L., Attardi, L. D., Verrijzer, C. P., Yokomori, K. and Tjian, R. (1994). Assembly of recombinant TFIID reveals different co-activator requirements for different transcriptional activators. *Cell* **79**, 93–105.

Choy, B. and Green, M. R. (1993). Eukaryotic activators function during multiple steps of pre-initiation complex assembly. *Nature* **366**, 531–536.

Clark, A. R. and Docherty, K. (1993). Negative regulation of transcription in eukaryotes. *Biochemical Journal* **296**, 521–541.

Courey, A. J. and Tjian, R. (1988). Analysis of Sp1 *in vivo* reveals multiple transcriptional domains including a novel glutamine-rich activation motif. *Cell* **55**, 887–898.

Cowell, I. G. and Hurst, H. C. (1994). Transcriptional repression by the human bZIP factor E4BP4: definition of a minimal repressor domain. *Nucleic Acids Research* **22**, 59–65.

Croston, G. E. and Kadonaga, J. T. (1993). Role of chromatin structure in the regulation of transcription by RNA polymerase II. *Current Opinion in Cell Biology* **5**, 417–423.

Drapkin, R., Merino, A. and Reinberg, D. (1993). Regulation of RNA polymerase II transcription. *Current Opinion in Cell Biology* **5**, 469–476.

Farkas, G., Gausz, J., Galloni, M., Renter, G., Gyurkovics, H. and Karch, F. (1994). The trithorax-like gene encodes the *Drosophila* GAGA factor. *Nature* **371**, 806–808.

Felsenfeld, G. (1992). Chromatin as an essential part of the transcriptional mechanism. *Nature* **355**, 219–224.

Franke, A., De Camillis, M., Zirk, D., Cheng, N., Brock, H. W. and Paro, R. (1992). Polycomb and polyhomeotic are constitutents of a multimeric complex in chromatin of *Drosophila melanogaster*. *EMBO Journal* **11**, 2941–2950.

Gerber, H.-P., Seipel, K., Georgiev, O., Hofferer, M., Hug, M., Rusioni, S. and Schaffner, W. (1994). Transcriptional activation modulated by homopolymeric glutamine and proline residues. *Science* **263**, 808–811.

Gill, G. and Ptashne, M. (1987). Mutants of GAL4 protein altered in an activation function. *Cell* **51**, 121–126.

Gill, G. and Ptashne, M. (1988). Negative effect of the transcriptional activator GAL4. *Nature* **334**, 721–724.

Goding, C. R. and O'Hare, P. (1989). Herpes simplex virus Vmw65-octamer binding protein interaction: a paradigm for combinatorial control of transcription. *Virology* **173**, 363–367.

Goodbourn, S. (1990). Negative regulation of transcriptional initiation in eukaryotes. *Biochimica et Biophysica Acta* **1032**, 53–77.

Goodrich, J. A. and Tjian, R. (1994). TBP–TAF complexes: selectivity factors for eukaryotic transcription. *Current Opinion in Cell Biology* **6**, 403–409.

Greenblatt, J., Nodwell, J. R. and Mason, S. W. (1993). Transcriptional anti-termination. *Nature* **364,** 401–406.

Guarente, L. (1988). USAs and enhancers: common mechanism of transcriptional activation in yeast and mammals. *Cell* **52**, 303–305.

Hagen, G., Muller, S., Beato, M. and Suske, G. (1994). Sp1-mediated transcriptional activation is repressed by Sp3. *EMBO Journal* **13**, 3843–3851.

Hahn, S. (1993a). Structure (?) and function of acidic transcription activators. *Cell* **72**, 481–483.

Hahn, S. (1993b). Efficency in activation. *Nature* **363**, 672–673.

Han, K. and Manley, J. L. (1993). Transcriptional repression by the *Drosophila* even skipped protein: definition of a minimal repressor domain. *Genes and Development* **7**, 491–503.

Herschbach, B. M. and Johnson, A. D. (1993). Transcriptional repression in eukaryotes. *Annual Review of Cell Biology,* **9**, 479–509.

Hollenberg, S. M. and Evans, R. M. (1988). Multiple and cooperative trans-activation domains of the human glucocorticoid receptor. *Cell* **55**, 899–906.

Hope, I. A. and Struhl, K. (1986). Functional dissection of a eukaryotic transcriptional activator, GCN4 of yeast. *Cell* **46**, 885–894.

Horey, R. and Carey, M. (1994). The role of activators in assembly of RNA polymerase II transcription complexes. *Current Opinion in Genetics and Development* **4**, 236–244.

Horikoshi, M., Carey, M. F., Kakidani, H. and Roeder, R.G. (1988a). Mechanism of action of a yeast activator: direct effect of GAL4 derivatives on mammalian TFIID promoter interactions. *Cell* **54**, 665–669.

Horikoshi, M., Hai, T., Lin, Y-S., Green, M. R. and Roeder, R. G. (1988b). Transcription factor ATF interacts with the TATA box factor to facilitate establishment of a preinitiation complex. *Cell* **54**, 1033–1042.

Inostroza, J. A., Mermelstein, F. H., Ha, I., Lane, W. S. and Reinberg, D. (1992). Dr1, a TATA-binding protein-associated phosphoprotein and inhibitor of class II gene transcription. *Cell* **70**, 477–489.

Johnston, S. A., Salmeron, J. M. and Pincher, S. E. (1987). Interaction of positive and negative regulatory proteins in the galactose regulon of yeast. *Cell* **50**, 143–146.

Joliot, V., Demma, M. and Prywes, R. (1995). Interaction with RAP74 subunit of TFIIF is required for transcriptional activation by serum response factor. *Nature* **373**, 632–635.

Kakidani, H. and Ptashne, M. (1988). Gal4 activates gene expression in mammalian cells. *Cell* **52**, 161–167.

Kakkis, E., Riggs, K. J., Gillespie, W. and Calame, K. (1989). A transcriptional repressor of C-myc. *Nature* **339**, 718–721.

Kim, Y.-J., Bjorklund, S., Li, Y., Sayre, M. H. and Kornberg, R. D. (1994). A multi-protein mediator of transcriptional activation and its interaction with the C-terminal domain of RNA polymerase II. *Cell* **77**, 599–608.

Kunzler, M., Braus, G. H., Georgiev, O., Seipel, K. and Schaffner, W. (1994). Functional differences between mammalian transcriptional activation domains at the yeast GAL1 promoter. *EMBO Journal* **13**, 641–645.

Latchman, D. S. (1995). *Gene Regulation, A Eukaryotic Perspective*, 2nd ed. London: Chapman and Hall.

Lee, J. W., Ryan, F., Swaffield, J. C., Johnston, S. A. and Moore, D. D. (1995). Interaction of thyroid-hormone receptor with a conserved transcriptional activator. *Nature* **374**, 91–94.

Leuther, K. K., Salmeron, J. M. and Johnston, S. A. (1993). Genetic evidence that an activation domain of GAL4 does not require acidity and may form a β sheet. *Cell* **72**, 575–585.

Lewin, B. (1994). Chromatin and gene expression: constant questions but changing answers. *Cell* **79**, 397–406.

Lillycrop, K. A., Dawson, S. J., Estridge, J. K., Gester, T., Matthias, P. and Latchman, D. S. (1994). Repression of a herpes simplex virus immediate-early promoter by the Oct-2 transcription factor is dependent upon an inhibitory region at the N-terminus of the protein. *Molecular and Cellular Biology* **14**, 7633–7642.

Lin., Y.-S., Carey, M., Ptashne, M. and Green, M. R. (1990). How different eukaryotic activators can co-operate promiscuously. *Nature* **345**, 359–361.

Lis, J. and Wu, C. (1993). Protein traffic on the heat shock promoter: parking, stalling and trucking along. *Cell* **74**, 1–4.

Mermod, N., O'Neil, E. A., Kelley, T. J. and Tjian, R. (1989). The proline-rich transcriptional activator of CTF/NF-1 is distinct from the replication and DNA binding domain. *Cell* **58**, 741–753.

Mitchell, P. J. and Tjian, R. (1989). Transcriptional regulation in mammalian cells by sequence specific DNA binding proteins. *Science* **245**, 371–378.

Ozer, J., Moore, P. A., Bolden, A. H., Lee, A., Rosen, C. A. and Lieberman, P. M. (1994). Molecular cloning of the small (α) sub-unit of human TFIIA reveals functions critical for activated transcription. *Genes and Development* **8**, 2324.

Paranjape, S. M., Kamakaka, R. T. and Kadonaga, J. T. (1994). Role of chromatin structure in the regulation of transcription by RNA polymerase II. *Annual Review of Biochemistry* **63**, 265–297.

Ptashne, M. (1988). How eukaryotic transcriptional activators work. *Nature* **335**, 683–689.

Roberts, S. G. E. and Green, M. R. (1994). Activator-induced conformational change in general transcription factor TFIIB. *Nature* **371**, 717–720.

Scafe, C., Chao, D., Lopes, J., Hirsch, J. P., Henry, S. and Young, R. A. (1990). RNA polymerase II C-terminal repeat influences response to transcriptional enhancer signals. *Nature* **347**, 491–494.

Seipel, K., Georgiev, O. and Schaffner, W. (1992). Different activation domains stimulate transcription from the remote ('enhancer') and proximal ('promoter') positions. *EMBO Journal* **11**, 4961–4968.

Sharp, P. A. (1991). TFIIB or not TFIIB? *Nature* **351**, 16–18.

Sigler, P. B. (1988). Acid blobs and negative noodles. *Nature* **333**, 210–212.

Spencer, C. A. and Groudine, M. (1990). Transcriptional elongation and eukaryotic gene regulation. *Oncogene* **5**, 777–785.

Suaren, J., Schmitz, J. and Horz, W. (1994). The transactivation domain of PHO4 is required for nucleosome disruption at the PHO5 promoter. *EMBO Journal* **13**, 4562–4586.

Swaffield, J. C., Melcher, K. and Johnston, S. A. (1995). A highly conserved ATPase protein as a mediator between acidic activation domains and the TATA-binding protein. *Nature* **374**, 88–91.

Tasset, D., Tora, D., Fromenthul, C., Scheer, E. and Chambon, P. (1990). Distinct classes of transcriptional activating domains function by different mechanisms. *Cell* **62**, 1177–1187.

Tjian, R. and Maniatis, T. (1994). Transcriptional activation: a complex puzzle with few easy pieces. *Cell* **77**, 5–8.

Travers, A. A. (1994). Chromatin structure and dynamics. *Bioessays* **16**, 657–662.

Wang, E. H. and Tjian, R. (1994). Promoter-selective transcriptional defect in cell cycle mutant ts13 rescued by hTAF$_{II}$ 250. *Science* **263**, 811–814.

White, R. J., Khoo, B. C.-E., Inostroza, J. A., Reinberg, D. and Jackson, S. P. (1994). Differential regulation of RNA polymerases I, II and III by the TBP-binding repressor Dr1. *Science* **266**, 448–450.

Xiao, H., Pearson, A., Coulombe, B., Truant, R., Zhang, S., Regier, J. L., Triezenberg, S. J., Reinberg, D., Flores, O., Ingles, C. J. and Greenblatt, J. (1994). Binding of basal transcription factor TFIIH to the acidic activation domains of VP16 and p53. *Molecular and Cellular Biology* **14**, 7013–7024.

Yankulov, K., Blau, J., Purton, T., Roberts, S. and Bentley, D.L. (1994). Transcriptional elongation by RNA polymerase II is stimulated by transactivators. *Cell* **77**, 749–759.

CHAPTER TEN

What regulates the regulators?

10.1 INTRODUCTION

As we discussed in previous chapters, transcription factors, in addition to their role in constitutive gene expression (Chapter 3), also play a central role in producing the induction of specific genes in response to particular stimuli (Chapter 4) and in the cell type-specific (Chapter 5) or developmentally regulated (Chapter 6) expression of other genes. The ability to bind to DNA (Chapter 8) and influence the rate of transcription either positively or negatively (Chapter 9) are clearly features of many transcription factors that regulate gene expression in response to specific stimuli or in specific cell types. Most importantly, however, such factors must also have their activity regulated such that they become active only in the appropriate cell type or in response to the appropriate stimulus, thereby producing the desired pattern of gene expression. The means by which this occurs have already been discussed in the appropriate chapters dealing with specific transcription factors. The aim of this chapter is to categorize the various means by which this regulation of transcription factor activity can be achieved and to illustrate these processes, both with examples that have already been discussed and with additional cases involving factors that have not so far been considered.

Two basic mechanisms by which the action of transcription factors can be regulated have been described. These involve either controlling the synthesis of the transcription factor so that it is made only when necessary (Figure 10.1a) or, alternatively, regulating the activity of

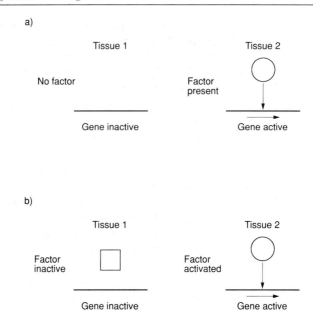

Figure 10.1 Gene activation mediated by the synthesis of a transcription factor only in a specific tissue (a) or its activation in a specific tissue (b).

the factor so that pre-existing protein becomes activated when required (Figure 10.1b). These two mechanisms will be discussed in turn.

10.2 REGULATION OF SYNTHESIS

10.2.1 Evidence for the regulated synthesis of transcription factors

Regulating the synthesis of transcription factors, such that they are made only when the genes they regulate are to be activated, is an obvious mechanism of ensuring that specific genes become activated only at the appropriate time and place. This mechanism is widely used, therefore, particularly for transcription factors that regulate the expression of cell type-specific or developmentally regulated genes. Thus, as discussed in Section 5.2.2, the octamer-binding protein Oct-2, which stimulates imunoglobulin gene enhancer activity in B cells, is detectable in these cells, but is absent from many other cell types, such as HeLa cells which do not express the immunoglobulin genes (Scheidereit *et al.*, 1987). Such cell type-specific expression of the Oct-

2 protein is paralleled by the presence of the Oct-2 mRNA in B cells and not in HeLa cells (Staudt *et al.*, 1988). Hence the cell type-specific synthesis of the Oct-2 protein is controlled by regulating the synthesis of the Oct-2 mRNA. Similarly, the MyoD transcription factor and its corresponding mRNA are detectable only in cells of the skeletal muscle lineage in which MyoD plays a critical role in regulating muscle-specific gene expression (Davis *et al.*, 1987; see Section 5.3.2).

In addition to its role in controlling cell type-specific gene expression, regulation of transcription factor synthesis is also widely used in the control of developmentally regulated gene expression. Thus numerous studies of the *Drosophila* homeobox transcription factors discussed in Section 6.2 using both immunofluorescence with specific antibodies and *in situ* hybridization, have revealed highly specific expression patterns for individual factors and the mRNAs which encode them indicating that their role in regulating gene expression in development is dependent, at least in part, on the regulation of their synthesis (review: Ingham, 1988; Figure 10.2). In agreement with this idea, alteration of the levels of the homeobox protein Bicoid within individual cells of the *Drosophila* embryo alters the phenotype of the cell to that characteristic of cells which normally contain the new level of the Bicoid protein (Driever and Nusslein-Volhard, 1988). Hence, in a number of cases, where a factor must be active in a particular cell type or at a specific point in development, this is achieved by the factor being present only in the particular cells where it is required. Clearly, such regulated synthesis of a specific transcription factor could be achieved by any of the

Figure 10.2 Localization of the Ftz protein in the *Drosophila* blastoderm embryo using a fluorescent antibody which reacts specifically with the protein. The anterior end of the embryo is to the left and the dorsal surface to the top of the photograph. Note the precise pattern of seven stripes of Ftz-expressing cells around the embryo.

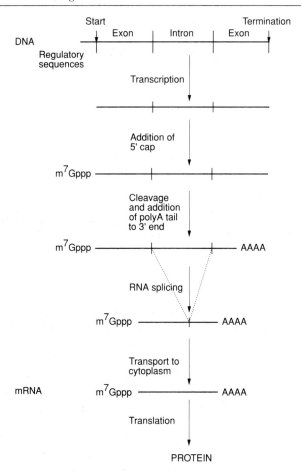

Figure 10.3 Potential regulatory stages in the expression of a gene encoding a transcription factor.

methods that are normally used to regulate the production of individual proteins, such as the regulation of gene transcription, RNA splicing or translation of the mRNA (Figure 10.3; for review of the levels at which gene regulation can occur, see Latchman (1995)). Several of these mechanisms of gene regulation are utilized in the case of individual transcription factors and these will be discussed in turn.

10.2.2 Regulation of transcription

As discussed above, a number of cases where the cell type-specific expression of a transcription factor is paralleled by the presence of its

Figure 10.4 Nuclear run-on assay of transcription in the nuclei of kidney and liver. Values indicate the degree of transcription of each gene in the two tissues. Note the enhanced transcription in the liver of the gene encoding the transcription factor C/EBP as well as of the genes encoding the liver-specific proteins transthyretin (TTR) and α_1-antitrypsin (α1AT). The positive control transfer RNA gene is, as expected, transcribed at equal levels in both tissues, whilst the negative control, pBR322 bacterial plasmid, does not show any transcription.

corresponding mRNA in the same cell type have now been described. In turn this cell type-specific expression of the transcription factor mRNA is likely to result from the regulated transcription of the gene encoding the transcription factor. Unfortunately, the low abundance of many transcription factors has precluded direct demonstration of the regulated transcription of the genes that encode them. This has been achieved, however, in the case of the CCAAT box-binding factor C/EBP (Section 3.3.3) which regulates the transcription of several different liver-specific genes such as transthyretin and α_1-antitrypsin. Thus, by using nuclear run-on assays to measure transcription of the gene encoding C/EBP directly, Xanthopoulos *et al.* (1989) were able to show that this gene is transcribed at high levels only in the liver, paralleling the presence of C/EBP itself and the mRNA encoding it at high levels only in this tissue (Figure 10.4). Hence the regulated transcription of the C/EBP gene in turn controls production of the corresponding protein which, in turn, directly controls the liver-specific transcription of other genes such as α_1-antitrypsin and transthyretin.

Interestingly, as well as being used to regulate the relative amounts of a particular factor produced by different tissues, transcriptional control can also be used to regulate factor levels within a specific cell type. Thus the levels of the liver-specific transcription factor DBP are highest in rat hepatocytes in the afternoon and evening, with the protein being undetectable in the morning. This fluctuation is produced by regulated transcription of the gene encoding DBP, which is highest in the early evening and undetectable in the morning whereas the C/EBP gene is transcribed at equal levels at all times. In turn the alterations in DBP level produced in this way produce similar diurnal fluctuations in the transcription of the albumin gene, which is dependent on DBP for its transcription (Wuarin and Schibler, 1990).

Although regulated transcription of the genes encoding the transcription factors themselves is likely, therefore, to constitute the predominant means of regulating their synthesis, it is clear that this process simply sets the problem of gene regulation one stage further back. Thus it will be necessary to have some means of regulating the specific transcription of the gene encoding the transcription factor itself, which in turn may require other transcription factors that are synthesized or are active only in that specific cell type. It is not surprising, therefore, that the synthesis of transcription factors is often modulated by post-transcriptional control mechanisms not requiring additional transcription factors. These mechanisms will now be discussed.

10.2.3 Regulation of RNA splicing

Numerous examples have now been described in eukaryotes where a single RNA species transcribed from a particular gene can be spliced in two or more different ways to yield different mRNAs encoding proteins with different properties (reviews: Latchman 1990, 1995; McKeown, 1992). This process is also used in several cases of genes encoding specific transcription factors, for example in the case of the *era-1* gene, which encodes a transcription factor that mediates the induction of gene expression in early embryonic cells in response to retinoic acid. In this case two alternative spliced mRNAs are produced, one of which encodes the active form of the molecule, whilst the other produces a protein lacking the homeobox region. As the homeobox mediates DNA binding by the intact protein (see Section 6.2.2), this truncated form of the protein is incapable of binding to DNA and activating gene expression (Larosa and Gudas, 1988). A similar use of alternative splicing to create mRNAs encoding proteins with and without the homeobox has also been reported for the Hox b6 (2.2) gene (Shen *et al.*, 1991).

Hence, in these cases where one of the two proteins encoded by the alternatively spliced mRNAs is inactive, alternative splicing can be used in the same way as the regulation of transcription in order to control the amount of functional protein produced.

Interestingly, however, unlike transcriptional regulation, alternative splicing can also be used to regulate the relative production of two distinct functional forms of a transcription factor which have different properties. Such alternative splicing is observed in both the CREB and CREM factors, which are encoded by two distinct genes. These two factors mediate the induction of specific genes in response to cyclic

AMP by binding to specific cyclic AMP response elements (CREs) in the promoters of inducible genes (reviews: Lalli and Sassone-Corsi, 1994; Sassone-Corsi, 1994). Although different products are produced from the CREB gene by alternative splicing (Yamamoto *et al.*, 1990), this process has been more intensively studied in the case of the distinct gene encoding the CREM factor and this will, therefore, be discussed in detail.

The CREM gene encodes a factor containing two glutamine-rich

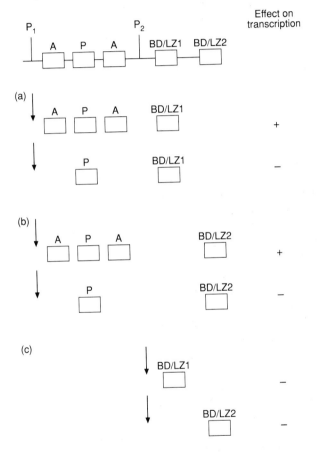

Figure 10.5 The CREM protein contains two transcriptional activation domains (A), a region containing a site for cyclic AMP-induced phosphorylation (P) (see section 10.3.4) and two DNA-binding domains containing a basic domain and leucine zipper (BD/LZ). After transcription from the P_1 promoter, alternative splicing can result in forms with or without the activation domains (a) or having either of the DNA-binding domains (b). In addition, cyclic AMP-inducible transcription from the P_2 promoter can produce forms containing only one or other of the DNA-binding domains but lacking the activation domains and the phosphorylated region (c). Arrows indicate the transcriptional start sites used in each case.

Figure 10.6 Gene activation by the activating forms of the CREM protein (A) can be inhibited by forms (I) which contain the DNA binding-domain (light shading) but lack the activation domain (heavy shading). They therefore bind to the CRE and prevent binding by the activating forms.

activation domains (see Section 9.2.2) and can, therefore, activate gene expression in response to cyclic AMP. Interestingly, however, alternative splicing produces distinct forms of the factor which lack these activation domains, although they retain the leucine zipper and basic DNA-binding domain (Figure 10.5a). These forms can therefore bind to DNA but cannot activate transcription, since they lack an activation domain. They therefore inhibit transcription by competing for binding to the CRE with the activating forms (Figure 10.6) (see Section 9.3.1 for a discussion of indirect repression of this type). Since the proportion of the activating and inhibitory forms of CREM varies in different cell types, the level of transcription directed by a CRE following cyclic AMP treatment will be different in these cells, depending on the precise balance between the activating and inhibitory forms.

As well as producing distinct forms with and without the activation domain, the CREM factor also undergoes alternative splicing in another manner. Thus two distinct exons in the CREM gene contain two distinct DNA-binding domains. Alternative splicing results in the proteins, which either do or do not contain the activation domains, also having one or other of the DNA-binding domains (Figure 10.5b). As the relative usage of the two DNA-binding domains is different in different cell types, this effect is likely to have biological significance, but its precise role is at present unclear.

The different forms of the CREM factor which have been discussed so far are all produced by alternative splicing of a single RNA transcript whose rate of production is unaffected by cyclic AMP. As discussed in Section 10.3.4, the ability of the CREB factor and the activating forms of CREM to switch on gene expression is stimulated post-translationally by their phosphorylation following cyclic AMP treatment, hence allowing them to switch on gene expression in response to cyclic AMP.

In contrast to such post-translational regulation, the CREM gene also contains a promoter which is activated in response to cyclic AMP. This promoter produces transcripts encoding short proteins which contain one or other of the DNA-binding domains but lack the activation domain and the phosphorylated region (Figure 10.5c). These proteins can, therefore, bind to the cyclic AMP response element and repress transcriptional activation by the activating forms, exactly as described above for the alternatively spliced forms lacking the activation domain. These forms are therefore known as ICERs (inducible cyclic AMP early repressors). As they are inducible by cyclic AMP, these forms are likely to play a key role in making the cyclic AMP response self-limiting. Thus, following cyclic AMP treatment, CREB and CREM will activate the expression of promoters containing a CRE, including that which produces the ICERs. The ICERs produced in this manner will then bind to the CRE and switch off the inducible genes by preventing the binding of CREB and CREM (Figure 10.7), thereby making the cyclic AMP response a transient one.

The regulation of cyclic AMP-inducible transcription by the CREB and CREM factor is, therefore, extraordinarily complex, with both alternative splicing in the CREB and CREM genes and the use of two different promoters in the case of the CREM gene. Indeed, this may seem a highly complex system to use simply to mediate a relatively simple processes like cyclic AMP inducibility. However, the recent findings that cyclic AMP-inducible transcription mediated by these proteins plays a critical role in processes as diverse as pituitary gland development and long-term memory (reviews: Lalli and Sassone-Corsi, 1994; Stevens, 1994) indicates that this complexity may be required to regulate these very complex biological phenomena properly.

The repressor molecules generated by alternative splicing of the CREB and CREM transcripts act as indirect repressors which passively prevent DNA binding by the positively acting forms containing a functional activation domain (see Section 9.3.1). Alternative splicing can also occur, however, in factors that contain a specific inhibitory domain and which can therefore function as direct repressors, interfering with the activity of the basal transcriptional complex (see Section 9.3.2). Thus, although it is transcribed in B cells and not in most other cell types (see Section 10.2.1), the gene encoding the Oct-2 transcription factor is also transcribed in neuronal cells. In neuronal cells, the Oct-2 RNA is spliced so that the protein it encodes does not contain the C-terminal activation domain that allows it to activate transcription. It does, however, retain the N-terminal inhibitory domain discussed in Section 9.3.2 as well as the DNA-binding domain, and can therefore act as a direct inhibitor of gene expression (Lillycrop and Latchman, 1992; Lillycrop *et al.*, 1994). In contrast in B cells, alternative splicing produces an mRNA which encodes a protein containing both the inhibitory domain and the stronger activation domain, and which therefore activates transcription (Figure 10.8). Hence, in this case, alternative splicing produces different forms of a factor in different cell types which have opposite effects on the activity of their target promoters.

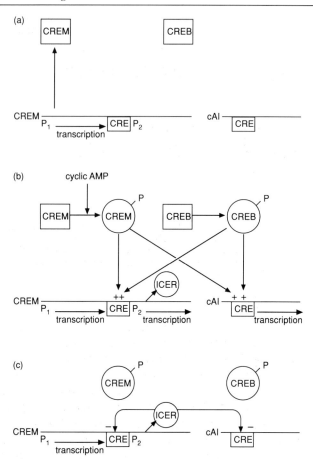

Figure 10.7 (a) In the absence of cyclic AMP, the CREM gene is transcribed from the P_1 promoter. However, neither the CREM produced in this way nor the CREB protein can activate transcription until they are activated post-translationally. (b) Following cyclic AMP treatment, the CREB and CREM proteins become activated post-translationally by phosphorylation. They therefore activate the cyclic AMP-inducible genes (cAI), which contain a cyclic AMP response element (CRE) in their promoters. In addition, they also activate the P_2 promoter of CREM, which also contains a CRE. (c) The ICER produced by the CREM P_2 promoter binds to the CREs and prevents activation by CREB and CREM, thereby repressing transcription.

Such alternative splicing is also seen in the case of another transcription factor containing an inhibitory domain, namely the thyroid hormone receptor (see Section 9.3.2). Thus, as discussed in Section 4.3.4, alternative splicing produces two forms of the receptor, one of which lacks the ligand-binding domain and, therefore, cannot bind thyroid hormone (Figure 4.20). Although it cannot, therefore, respond to thyroid hormone, this α2 form of the protein still contains the DNA-binding domain

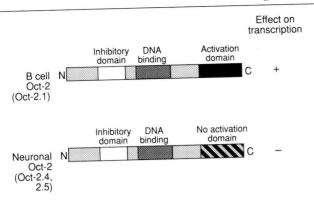

Effect on
transcription

Figure 10.8 In B lymphocytes the predominant form of Oct-2 (Oct-2.1) contains the C-terminal activation domain as well as the DNA-binding domain and an inhibitory domain. As the activation domain overcomes the effect of the inhibitory domain, this form is able to activate transcription. In contrast, the predominant neuronal forms of Oct-2 (Oct-2.4 and -2.5) contain different C-terminal regions and lack the activation domain. As they retain both the inhibitory domain and the DNA-binding domain, however, they can bind to specific DNA-binding sites and inhibit gene expression.

and can, therefore, bind to the specific binding site for the receptor in hormone-responsive genes. By doing so, it acts as a dominant repressor of gene activation mediated by the normal receptor in response to hormone binding. Hence these two alternatively spliced forms of the transcription factor, which are made in different amounts in different tissues, mediate opposing effects on thyroid hormone-dependent gene expression.

As well as affecting the activation domain, alternative splicing can affect the ability to heterodimerize with another protein. This is seen in the case of the *Drosophila* POU factors, I-POU and twin of I-POU (tI-POU) (Treacy *et al.*, 1992). Thus these two factors are generated by alternative splicing, resulting in their being identical except that tI-POU contains two additional amino acids in the region immediately N-terminal to the POU homeodomain (see Section 6.3.2 for a discussion of POU proteins). Interestingly, I-POU cannot bind to DNA but can interact with another POU protein, CFIa, preventing it from binding to DNA and activating transcription. In contrast, the two additional amino acids in tI-POU allow it to bind to DNA but prevent it interacting with CF1a. Hence, in this case, alternative splicing generates two proteins not only with different DNA-binding specificities but also different abilities to interact with other proteins. The examples of alternative splicing discussed above thus illustrate the potential of this process in generating different forms of a particular transcription factor which, because of differences in the regions that mediate DNA binding, transcriptional activation or protein–protein interaction, have different properties that result in differences in their effects on gene expression.

10.2.4 Regulation of translation

The final stage in the expression of a gene is the translation of its corresponding mRNA into protein. In theory, therefore, the regulation of synthesis of a particular transcription factor could be achieved by producing its mRNA in all cell types but translating it into active protein only in the particular cell type where it was required. However, the observed parallels between the cell type-specific expression of a particular transcription factor and the cell type-specific expression of its corresponding mRNA discussed above (Section 10.2.1) indicate that this cannot be the case for the majority of transcription factors. None the less this mechanism is used to control the synthesis of at least one transcription factor in yeast.

Thus the yeast GCN4 transcription factor controls the activation of several genes in response to amino acid starvation and the factor itself is synthesized in increased amounts following such starvation, allowing it to mediate this effect. This increased synthesis of GCN4 following amino acid starvation is mediated via increased translation of pre-existing GCN4 mRNA (review: Altmann and Trachsel, 1993). This translational regulation is dependent upon short sequences within the 5′ untranslated region of the GCN4 mRNA, upstream of the start point for translation of the GCN4 protein.

Most interestingly, such sequences are capable of being translated to produce short peptides of two or three amino acids (Figure 10.9). Under conditions where amino acids are plentiful, these short peptides are synthesized and the ribosome fails to reinitiate at the start point for GCN4 production, resulting in this protein not being synthesized. Following amino acid starvation, however, the production of the small peptides is suppressed and the production of GCN4 is correspondingly enhanced. Hence this mechanism ensures that GCN4 is synthesized

Figure 10.9 Presence of short open reading frames capable of producing small peptides in the 5′ untranslated region of the yeast GCN4 RNA. Translation of the RNA to produce these small proteins suppresses translation of the GCN4 protein. The position of the methionine residue beginning each of the small peptides is indicated, together with the number of additional amino acids incorporated before a stop codon is reached.

only in response to amino acid starvation and then activates the genes encoding the enzymes required for the biosynthetic pathways necessary to make good this deficiency.

Interestingly, the use of distinct translational start sites is also seen in the case of a transcription factor expressed in the mammalian liver. In this case, however, the two start sites of translation result in two different forms of the protein. One of these, known as liver activator protein (LAP), contains an activation domain as well as a basic DNA-binding domain and leucine zipper. The other is produced by translational initiation from a downstream start site and therefore lacks the activation domain, although it retains both the basic domain and the leucine zipper (Figure 10.10). This protein is, therefore, known as liver inhibitory protein (LIP), since it can bind to the same sites as LAP and inhibit gene activation (Descombes and Schibler, 1991).

Hence, as with the regulation of splicing, the regulation of translation can be used to control the amount of an active factor which is produced as well as to regulate the balance between two functionally antagonistic factors encoded by the same gene.

10.2.5 Role of regulated synthesis

Regulating the synthesis of a transcription factor constitutes a metabolically inexpensive way of controlling its activity. Thus, in situations where the activity of a particular factor is not required, no energy is expended on making it in an inactive form. Such regulation probably takes place predominantly at the level of transcription, so that no energy is expended on the production of an RNA, its splicing, transport etc. However, even in cases where regulation occurs at later stages such

Figure 10.10 The use of different translational initiation codons (vertical arrows) produces the LAP protein, which possesses an activation domain, and the LIP protein, which lacks this domain and therefore inhibits gene activation by LAP.

as splicing or translation, the system is relatively efficient in terms of energy usage, since the step in gene expression that requires the most energy is the final one of translation.

In view of its metabolic efficiency, it is not surprising, therefore, that the regulation of their synthesis is widely used to control the activity of factors which mediate cell type-specific gene regulation, where differences in the activity of a given factor in different cell types are maintained for long periods of time. Similarly, alternative splicing or use of different translational initiation codons is used to produce different forms of the same factor which often have antagonistic effects on gene expression.

The regulation of factor activity by regulating its synthesis does suffer, however, from the defect that a change in the level of activity of a factor which is controlled purely by a change in its actual amount can take some time to occur. Thus, in response to a signal that induces new transcription of the gene encoding a particular factor, it is necessary to go through all the stages illustrated in Figure 10.3 before the production of active factor which is capable of activating the expression of other genes in response to the inducing signal. It is not surprising, therefore, that, although some factors such as GCN4 which mediate inducible gene expression are regulated by the regulation of their synthesis, the majority of such factors are regulated by post-translational mechanisms which activate pre-existing transcription factor protein in response to the inducing signal. Thus, although mechanisms of this type are metabolically expensive in that they require the synthesis of the factor in situations where it is not required, they have the necessary rapid response time required for the regulation of inducible gene expression. Moreover, unlike transcriptional regulation, they constitute an independent method of gene regulation rather than requiring the activation of other transcription factors in order to activate the transcription of the gene encoding the factor itself.

10.3 REGULATION OF ACTIVITY

10.3.1 Evidence for the regulated activity of transcription factors

In a number of cases it has been shown that a particular transcription factor pre-exists in an inactive form prior to its activation and the consequent switching on of the genes that depend on it for their activity. Thus, whilst the activation of heat-inducible genes by increased temperature is dependent on the activity of the heat-shock transcription factor (HSF), this induction can be achieved in the presence of the protein synthesis inhibitor cycloheximide (Zimarino and Wu, 1987). Hence this process cannot be dependent on the synthesis of HSF in response to heat but rather must depend on the heat-induced activation of pre-existing inactive HSF (see Section 4.2.2). Similarly, as discussed in Section 9.3.1, the yeast GAL4 transcription factor pre-exists in cells prior to galactose treatment, which activates it by causing the dissociation of the inhibitory GAL80 protein.

Although, for the reasons discussed above (Section 10.2.4), the activation of pre-existing transcription factors is predominantly used to modulate transcription factors involved in controlling inducible rather than cell type-specific gene expression, it has also been reported for factors involved in regulating cell type-specific gene expression. Thus, as discussed in Section 5.2.2, like Oct-2, the transcription factor NFκB (which is a heterodimer of two subunits, p50 and p65) plays an important role in the B cell-specific expression of the immunoglobulin κ gene (reviews: Blank *et al.*, 1992; Grimm and Baeurele, 1993). Unlike Oct-2, however, both subunits of NFκB are expressed in a wide variety of cell types and the factor is present in an inactive form both in pre-B cells and in a wide variety of other cell types, such as T cells and HeLa cells, which do not express the immunoglobulin genes. This pre-existing form of NFκB can be activated by treatment with substances such as lipopolysaccharides or phorbol esters. These treatments, therefore, result in the activation of the immunoglobulin κ gene in pre-B cells, which do not normally express it, and in the expression of other NFκB-dependent genes such as the interleukin-2α receptor in T cells. This mechanism, in which pre-existing NFκB becomes activated both during B cell differentiation and by agents such as phorbol esters which activate T cells, therefore allows NFκB to play a dual role both in B cell-specific gene expression and in the expression of particular genes in response to T cell activation by various agents (Figure 10.11). This effect would otherwise require a complex pattern of regulation in which NFκB was synthesized both in response to B cell maturation and to agents that activate T cells.

Hence modulating the activity of a transcription factor represents a rapid and flexible means of activating a particular factor. Moreover, unlike transcriptional control, such mechanisms allow a direct linkage between the inducing stimulus and the activation of the factor, rather than requiring the regulated activity of other transcription factors which in turn activate transcription of the gene encoding the regulated factor. Hence they represent a highly efficient means of allowing specific cellular signalling pathways to produce changes in cellular transcription factor activity and affect gene expression (reviews: Karin, 1992; Karin and Smeal, 1992; Hill and Treisman, 1995).

Figure 10.11 Activation of NFκB during B cell differentiation or by agents such as PMA which activate T cells allows it to activate expression of the immunoglobulin κ-chain gene in B cells and the interleukin 2 receptor gene in activated T cells.

In the most extreme example of the linkage between signalling pathways and transcription factors, the signalling molecule and the transcription factor are identical. Thus, in response to microbial infection, mammalian neutrophils secrete the protein lactoferrin into the medium. It has recently been shown that lactoferrin can be taken up by other cells of the immune system. It then enters the nucleus of the cells and binds to specific DNA sequences, activating genes whose protein products are required for the cells to neutralize the microbial infection (He and Furmanski, 1995). Hence, in this case, the signalling factor and the transcription factor are the same protein (for discussion see Baeurele (1995)). In most cases, however, the signalling molecule acts indirectly to produce a change in the activity of a distinct transcription factor which pre-existed within the cell in an inactive form prior to exposure to the signal. Three basic means by which such mechanisms can regulate factor activity have been described (Figure 10.12) and these will be discussed in turn.

10.3.2 Regulation by protein–ligand binding

As discussed above, one of the principal advantages of regulating the activity of a factor in response to an inducing stimulus is that it allows a direct interaction between the inducing stimulus and the activation of the factor, ensuring a rapid response. The simplest method for this is for

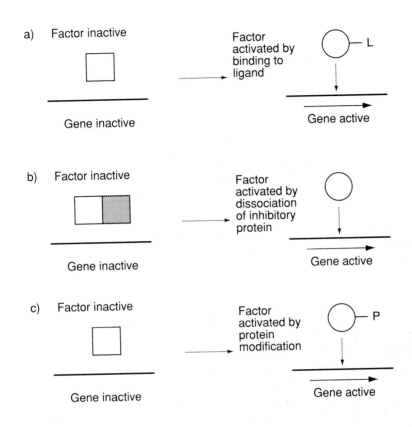

Figure 10.12 Methods of activating a transcription factor in response to an inducing stimulus. This can occur (a) by a ligand-mediated conformational change, (b) by removal of an inhibitory protein or (c) by a modification to the protein such as phosphorylation.

an inducing ligand to bind to the transcription factor and alter its structure so that it becomes activated (Figure 10.12a).

An example of this effect is seen in the case of the ACE1 factor, which mediates the induction of the yeast metallothionein gene in response to copper. In this case, the transcription factor undergoes a major conformational change in the presence of copper which converts it to an active form that is able to bind to its appropriate binding sites in the metallothionein gene promoter and activate transcription (Figure 10.13; review: Thiele, 1992).

A similar example of such ligand-induced regulation, which occurs in mammalian cells, is seen in the case of the thyroid hormone receptor. Thus, as discussed in Section 4.3.4 this receptor binds to DNA in the

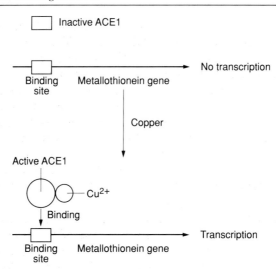

Figure 10.13 Activation of the ACE1 factor in response to copper results in transcription of the metallothionein gene.

absence of thyroid hormone and inhibits gene expression via a specific inhibitory domain. Upon binding of thyroid hormone, the receptor undergoes a conformational change which exposes its activation domain and allows it to activate transcription (see Figure 4.17). The importance of this conversion from repressor to activator is seen in the case of mutant forms of the thyroid hormone receptor which cannot undergo this conformational change because they do not bind thyroid hormone. This is observed not only in the v-*erbA* oncogene, as discussed in Section 7.2.2, but also in patients with generalized thyroid hormone resistance. Thus these patients have been shown to produce forms of the receptor which can repress gene expression but which cannot activate genes in response to thyroid hormone. Most interestingly, the presence of these dominant negative forms of the receptor results in impairment of physical and mental development which is much more severe than that observed if the receptor is absent completely (Baniahmad *et al.*, 1992).

Although the receptors for steroid hormones such as the glucocorticoids belong to the same family of receptors as the thyroid hormone receptor, they are regulated by their ligand in a somewhat different manner. Thus, prior to hormone addition, the glucocorticoid receptor is prevented from binding to DNA by its association with the 90 kDa heat-shock protein (hsp90) (see Section 4.3.2). Steroid treatment induces a conformational change in the receptor, resulting in its dissociation from hsp90. This step is essential for DNA binding by the receptors which, although inherently able to bind to DNA, cannot do so until they are released from the complex with hsp90 (see Figure 4.10) (review: Pratt, 1993; Smith and Toft, 1993). Most interestingly, the association of hsp90 with the glucocorticoid receptor occurs via the C-terminal region of the receptor, which also contains the steroid-binding domain. It has been suggested, therefore, that by associating with the C-terminal region of the receptor hsp90 masks adjacent domains whose activity is necessary for gene activation by the receptor, for example those involved in receptor dimerization or subsequent DNA binding, thereby preventing DNA binding from occurring. Following steroid treatment, however, the steroid binds to the C-terminus of the receptor, displacing hsp90 and thereby unmasking these domains and allowing DNA binding to occur (Figure 10.14).

Hence activation of the steroid receptors involves a ligand-induced conformational change which, unlike the case of the thyroid hormone receptor, results in the dissociation of an inhibitory protein. Paradoxically, however, it appears that prior association with hsp90 may also be necessary for the eventual activation of the receptors by hormone. Thus the glucocorticoid receptor binds steroid with high affinity when it is

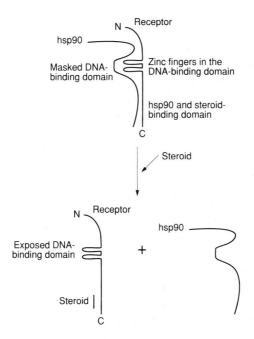

Figure 10.14 Interaction of hsp90 and the glucocorticoid receptor. Hsp90 binds to the receptor via the C-terminal region of the receptor, which also binds steroid and may mask regions of the receptor necessary for dimerization or DNA binding. When steroid is added it binds to the receptor at the C-terminus, displacing hsp90 and exposing the masked regions.

associated with hsp90 and this ability is lost if the hsp90 is artifically dissociated from the receptor by exposure to high temperature or specific ionic conditions (Bresnick *et al.*, 1989). This suggests that association with hsp90, as well as preventing activation of the receptors prior to steroid addition, may also alter the structure of the receptor so as to facilitate its activation when steroid is added. This could be achieved by hsp90 altering the conformation of the receptor so as to increase its ability to bind steroid. Hence, in this case, hsp90 is acting not only to maintain the transcription factor in an inactive form prior to ligand addition but also to increase the ability of the factor to become activated when ligand is added.

The association and subsequent steroid-induced dissociation of the receptors from hsp90, therefore, appears to play an important role in their activation. As discussed in Section 4.3.2, however, steroid treatment also has an additional effect on the receptors, which is essential for transcriptional activation by free receptor. Thus 'domain swapping' experiments of the type described in Section 2.3.3 have identified transcriptional activation domains in both the glucocorticoid and oestrogen receptors which are capable of stimulating transcription only following hormone addition. Hence the steroid-induced activation of the receptors appears to involve a conformational change in the receptor which occurs upon steroid binding and which unmasks a steroid-dependent activation domain in the receptor. Activation of the receptors by steroid, therefore, involves both ligand-induced conformational changes, analogous to those produced in the ACE1 factor by copper and in the thyroid hormone receptor by thyroid hormone, as well as the dissociation of an inhibitory protein. The various effects of steroids on the steroid receptors are summarized in Figure 10.15, whilst other examples of the

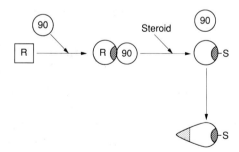

Figure 10.15 Interaction of the receptor (R) with hsp90 and steroid. Note that prior interaction with hsp90 may be necessary to facilitate receptor activation, possibly by enhancing its capacity to bind steroid (shaded region). Subsequent treatment with steroid results in the dissociation of hsp90 and also activates a steroid-dependent activation domain (dotted region).

regulation of transcription factors by protein–protein interactions are discussed in the next section.

10.3.3 Regulation by protein–protein interactions

As described above, the glucocorticoid receptor is regulated by its inter-action with hsp90, which prevents it binding to DNA and activating transcription in the absence of steroid hormone. A similar mechanism is used in the case of the NFκB factor which, as discussed above, only activates transcription in mature B cells or in other cell types following treatment with agents such as lipopolysaccharides or phorbol esters. In agreement with this, no active form of NFκB capable of binding to DNA can be detected in DNA mobility shift assays (see Section 2.2.1) using either cytoplasmic or nuclear extracts prepared from pre-B cells or non-B cell types. Interestingly, however, such activity can be detected in the cytoplasm but not the nucleus of such cells following denaturation and subsequent renaturation of the proteins in the extract. Hence NFκB exists in the cytoplasm of pre-B cells and other cell types in an inactive form which is complexed with another protein known as IκB that inhibits its activity (reviews: Beg and Baldwin, 1993; Gilmore and Morin, 1993; Thanos and Maniatis, 1995). The release of NFκB from IκB by the denaturation–renaturation treatment therefore results in the appearance of active NFκB capable of binding to DNA (Figure 10.16a).

These findings suggested, therefore, that treatments with substances such as lipopolysaccharides or phorbol esters do not activate NFκB by interacting directly with it in a manner analogous to the activation of the ACE1 factor by copper. Rather, they are likely to produce the dissociation of NFκB from IκB, resulting in its activation. In agreement with this idea, phorbol ester treatment of cells prior to their fractiona-tion eliminated the latent NFκB activity in the cytoplasm and resulted in the appearance of active NFκB in the nucleus (Figure 10.16b). These substances act, therefore, by releasing NFκB from IκB, allowing it to move to the nucleus where it can bind to DNA and activate gene expression. Hence this constitutes an example of the activation of a factor by the dissociation of an inhibitory protein (Figure 10.12b).

Such a mechanism is used to regulate the activity of many different transcription factors. Thus, apart from the NFκB–IκB and glucocorticoid receptor–hsp90 interactions, other examples of inhibitory interactions discussed in previous chapters include those between DNA binding helix–loop–helix proteins and Id (Section 5.3.2), Fos–Jun and the gluco-corticoid receptor (Section 4.3.4), and p53 and the MDM2 protein

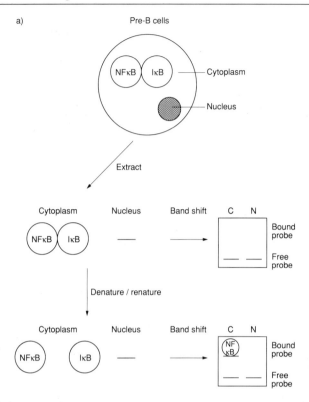

(Section 7.3.2). Hence inhibitory interactions of this type are widely used to regulate the activity of specific transcription factors.

In contrast, other transcription factors may be inactive alone and may need to complex with a second factor to be active. This is seen in the case of the Fos protein, which cannot bind to DNA without first forming a heterodimer with the Jun protein (see Section 7.2.1). A similar mechanism also operates in the case of the Myc factor, which cannot bind to DNA except as a complex with the Max protein (see Section 7.2.4). Hence protein–protein interactions between transcription factors can result in either inhibition or stimulation of their activity. The need for Fos and Myc to interact with another factor prior to DNA binding arises from their inability to form a homodimer, coupled with the need for factors of this type to bind to DNA as dimers. Hence they need to form heterodimers with another factor prior to DNA binding (see Section 8.4 for further discussion).

Even in the case of factors such as Jun, which can form DNA-binding homodimers, the formation of heterodimers with another factor offers the potential to produce a dimer with properties distinct from those of either homodimer. Thus the Jun homodimer can bind strongly to AP1

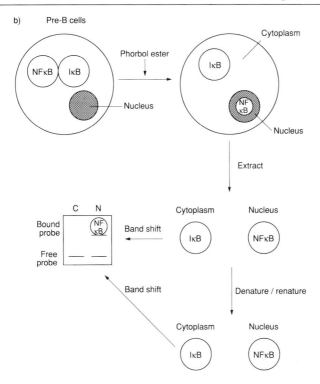

Figure 10.16 Regulation of NFκB. (a) In pre-B cells NFκB is located in the cytoplasm in an inactive form which is complexed to IκB. DNA mobility band shift assays do not, therefore, detect active NFκB. If a cytoplasmic extract is first denatured and renatured, however, active NFκB will be released from IκB and will be detected in a subsequent band shift assay. (b) In mature B cells, NFκB has been released from IκB and is present in the nucleus in an active DNA-binding form. It can therefore be detected in a DNA mobility shift assay without a denaturation–renaturation step, which has no effect on the binding activity.

sites but only weakly to the cyclic AMP response element (CRE). In contrast, a heterodimer of Jun and the CREB factor binds strongly to a CRE and more weakly to an AP1 site. Heterodimerization can, therefore, represent a means of producing multi-protein factors with unique properties different from that of either protein partner alone (reviews: Jones, 1990; Lamb and McKnight, 1991).

Hence, as well as stimulating or inhibiting the activity of a particular factor, the interaction with another factor can also alter its properties, directing it to specific DNA-binding sites to which it would not normally bind. This effect is also seen in the yeast mating-type system where, as discussed in Section 5.4.3, the α2 repressor is directed to different DNA-binding sites in different genes following its interaction with either the a1 or MCMI proteins. Similarly, as discussed in Section 6.2.3, the

Drosophila extradenticle protein changes the DNA-binding specificity of the Ubx protein so that it binds to certain DNA-binding sites with high affinity in the presence of extradenticle and with low affinity in its absence.

Although several examples of one transcription factor altering the DNA-binding specificity of another have thus been defined, such protein–protein interactions can also change the specificity of a transcription factor in at least one other way. This is seen in the case of the *Drosophila* dorsal protein, which is related to the mammalian NFκB factors. Thus this factor is capable of both activating and repressing specific genes. Such an ability is not due, for example, to the production of different forms by alternative splicing, since both activation and repression take place in the same cell type. Rather, it appears to depend on the existence of a DNA sequence (the ventral repression element (VRE)) adjacent to the dorsal-binding site in genes such as zen, which are repressed by dorsal, whereas the VRE sequence is absent in genes such as twist, which are activated by dorsal.

It has recently been shown (Lehming *et al.*, 1994) that DSP1 (dorsal switch protein), a member of the HMG family of transcription factors (see Section 8.5), binds to the VRE and interacts with the dorsal protein, changing it from an activator to a repressor. Hence in genes such as twist, where DSP1 cannot bind, dorsal activates expression, whereas in genes such as zen, which DSP1 can bind, dorsal represses expression (Figure 10.17). Interestingly, a similar negative element is associated with the NFκB-binding site in the mammalian β-interferon promoter, and the *Drosophila* DSP1 protein can similarly switch NFκB from activator to repressor when DSP1 is artifically expressed in mammalian cells. This mechanism may thus not be confined to *Drosophila* and a mammalian homologue of DSP1 may regulate NFκB activity in a similar manner (for discussion see Thanos and Maniatis (1995)).

Figure 10.17 The interaction of DSP1 bound at the ventral repression element (VRE) with the dorsal protein bound at its adjacent binding site (DBS) in the zen promoter results in dorsal acting as a repressor of transcription, whereas in the absence of binding sites for DSP1, as in the twist promoter, it acts as an activator.

Protein–protein interactions between different factors can thus either stimulate or inhibit their activity, or alter that activity, either in terms of DNA-binding specificity or even from activator to repressor. It is likely that the wide variety of protein–protein interactions and their diverse effects allow the relatively small number of transcription factors which exist to produce the complex patterns of gene expression that are required in normal development and differentiation.

10.3.4 Regulation by protein modification

Many transcription factors are modified extensively following translation by the addition of 0-linked monosaccharide residues (Jackson and Tjian, 1988) or by phosphorylation (reviews: Hunter and Karin, 1992; Hill and Treisman, 1995). Such modifications represent obvious targets for agents that induce gene activation. Thus, such agents could act by altering the activity of a modifying enzyme, such as a kinase. In turn this enzyme would modify the transcription factor, resulting in its activation and providing a simple and direct means of activating a particular factor in response to a specific signal (Figure 10.12c).

The most direct example of such an effect is seen in the case of gene activation by the interferons α and γ. Thus these molecules bind to cell surface receptors, which are associated with factors having tyrosine kinase activity. The binding of interferon to the receptor stimulates the kinase activity and results in the phosphorylation of transcription factors known as STATs. In turn this results in the movement of STAT

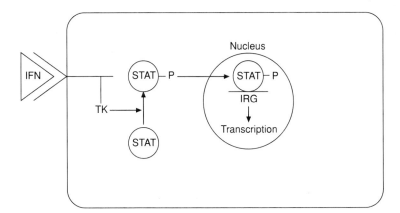

Figure 10.18 Binding of interferon (IFN) to its receptor results in activation of an associated tyrosine kinase (TK) activity, leading to phosphorylation of the STAT transcription factor and allowing it to move to the nucleus and stimulate interferon-responsive genes (IRG).

proteins to the nucleus, where they bind to DNA and activate interferon-responsive genes (Figure 10.18) (reviews: Darnell *et al.*, 1994; Ihle and Kerr, 1995).

Another example of this type is provided by the CREB factor, which mediates the induction of specific genes in response to cyclic AMP treatment (see Section 10.2.3). Thus treatment of cells with cyclic AMP is known to stimulate the activity of the protein kinase A enzyme (Nigg *et al.*, 1985) and in turn this enzyme phosphorylates CREB on a serine residue at position 133 in the molecule (reviews: Lalli and Sassone-Corsi, 1994; Sassone-Corsi, 1994) A similar phosphorylation on serine also occurs following cyclic AMP treatment in the case of the CREM factor (see Section 10.2.3). In both CREB and CREM, the region that is phosphorylated is located between the two glutamine-rich activation domains (see Figure 10.5). Phosphorylation stimulates the activity of the adjacent activation domains, resulting in the induction of cyclic AMP-responsive genes to which CREB or CREM has bound (Figure 10.19).

Hence, in this case, the activation of a specific enzyme by the inducing agent directly stimulates the ability of the transcription factor to activate transcription and hence results in the activation of cyclic AMP inducible genes. Similarly, the phosphorylation of the heat-shock factor (HSF) following exposure of cells to raised temperature increases the activity of its activation domain, leading to increased transcription of heat-inducible genes (see Section 4.2.2). In contrast, phosphorylation of the serum response factor (SRF), which mediates the induction of several mammalian genes in response to growth factors or serum addition, increases its ability to bind to DNA rather than directly increasing the activity of its activation domain (Manak *et al.*, 1990).

Figure 10.19 Activation of the CREB factor by cyclic AMP-induced phosphorylation. The ability of DNA-bound CREB to activate transcription is produced by the cyclic AMP-dependent activation of protein kinase A, which phosphorylates the CREB protein, resulting in its activation.

Interestingly, SRF normally binds to DNA in association with an accessory protein $p62^{TCF}$. The ability of $p62^{TCF}$ to associate with SRF is itself stimulated by phosphorylation (Gille *et al.*, 1992). In contrast, as discussed in Section 7.3.3, the association of the Rb-1 protein with other transcription factors, such as E2F and MyoD, is dependent upon the dephosphorylation of Rb-1. Hence the phosphorylation state of a transcription factor can control its ability to associate with other factors and regulate their activity, as well as its ability to enter the nucleus, bind to DNA or stimulate transcription.

The effect of phosphorylation on protein–protein interactions is also involved in the dissociation of NFκB and its associated inhibitory protein IκB, which was discussed above (Section 10.3.3). In this case, however, the target for phosphorylation appears to be the inhibitory protein IκB rather than the potentially active transcription factor itself. Thus, following treatment with phorbol esters which activate protein kinase C, IκB becomes phosphorylated. Such phosphorylation results in the dissociation of the NFκB–IκB complex and targets IκB for rapid degradation (Palombella *et al.*, 1994; Brown *et al.*, 1995). This breakdown of the complex results in NFκB being free to move to the nucleus and activate transcription (reviews: Beg and Baldwin, 1993; Grimm and Baeurele 1993; Thanos and Maniatis, 1995). Hence in this case, as before, the inducing agent has a direct effect on the activity of a kinase enzyme but the resulting phosphorylation inactivates an inhibitory factor rather than stimulating an activating factor. This example, therefore, involves a combination of two of the post-translational activation mechanisms we have discussed, namely protein modification and dissociation of an inhibitory protein.

In addition to its activation of NFκB, treatment with phorbol esters also results in the increased expression of several cellular genes which contain specific binding sites for the transcription factor AP1. As discussed in Section 7.2.1, this transcription factor in fact consists of a complex mixture of proteins, including the proto-oncogene products Fos and Jun. Following treatment of cells with phorbol esters, the ability of Jun to bind to AP1 sites in DNA is stimulated and this effect, together with the increased levels of Fos and Jun produced by phorbol ester treatment, results in the increased transcription of phorbol ester-inducible genes. As with the activation of NFκB, phorbol esters appear to increase DNA binding of Jun by activating protein kinase C. Paradoxically, however, it has been shown (Boyle *et al.*, 1991) that the increased DNA-binding ability of Jun following phorbol ester treatment is mediated by its dephosphorylation at three specific sites located immediately adjacent to the basic DNA-binding domain, indicating that protein kinase C acts by stimulating a phosphatase enzyme which in turn dephosphorylates Jun (Figure 10.20).

Figure 10.20 (a) Activation of Jun binding to DNA by dephosphorylation. The dephosphorylation of Jun protein following PMA treatment increases its ability to bind to AP1 sites and activate PMA-responsive genes. This is likely to be mediated via the PMA-dependent activation of protein kinase C, which in turn phosphorylates a phosphatase enzyme, allowing it to dephosphorylate Jun. (b) Position in the Jun protein of the two serine (S) and one threonine (T) residues, which are dephosphorylated in response to PMA. Note the close proximity to the basic domain (shaded) which mediates DNA binding. The positions of the trans-activation domain and leucine zipper are also indicated.

Such an inhibitory effect of phosphorylation on the activity of a transcription factor is not unique to the Jun protein, a similar effect of phosphorylation in reducing DNA-binding activity having also been observed in the Myb proto-oncogene protein, which was discussed in Section 7.2.3 (Luscher *et al.*, 1990). Moreover, DNA-binding ability is not the only target for such inhibitory effects of phosphorylation. Thus phosphorylation of the Bicoid protein reduces its ability to activate transcription without affecting its DNA-binding activity, presumably by inhibiting the activity of its activation domain (Ronchi *et al.*, 1993).

Hence protein modification by phosphorylation can have a wide variety of effects on transcription factors, either stimulating or inhibiting their activity and acting via a direct effect on the ability of the factor to enter the nucleus, bind to DNA, associate with another protein or activate transcription, or by an indirect effect affecting the activity of an inhibitory protein. The directness and rapidity of this means of transcription factor activation suggests that the other forms of modification that have been observed for specific transcription factors, such as glycosylation (Jackson and Tjian, 1988), are also likely to be the targets for regulatory processes.

10.3.5 Role of regulated activity

In addition to its ability to produce a very rapid activation of gene expression, modification of the activity of a pre-existing protein also allows specific targets for modification to be used in different cases. Thus, in the case of phosphorylation discussed above (Section 10.3.4), we have seen how in different cases a single process can increase either the DNA-binding ability or the trans-activation ability of specific factors, or affect their ability to associate with another protein..

Clearly, therefore, post-translational mechanisms for activating pre-existing protein could be used independently to stimulate either the DNA binding or the transcriptional activation activities of a single factor in different situations within a complex regulatory pathway. Indeed, such a combination of mechanisms is actually used to regulate the activity of the yeast GAL4 transcription factor. Thus, as discussed in Section 9.3.1, the activation of transcription of galactose-inducible genes by GAL4 is mediated by the galactose-induced dissociation of the inhibitory GAL80 protein, which exposes the activation domain of DNA-bound GAL4. Interestingly, however, this effect occurs only when the cells are grown in the presence of glycerol as the main carbon source. By contrast, however, in the presence of glucose, GAL4 does not bind to DNA and the addition of galactose has no effect (Giniger *et al.*, 1985). Hence, by having a system in which glucose modulates the DNA binding of the factor and galactose modulates the activation of bound factor, it is possible for glucose to inhibit the stimulatory effect of galactose. This ensures that the enzymes required for galactose metabolism are induced only in the presence of glycerol and not in the presence of the preferred nutrient, glucose (Figure 10.21).

Such a system, in which two different activities of a single factor are

Figure 10.21 Effects of glucose and galactose on GAL4 activity. Note that, whilst galactose stimulates the ability of DNA-bound GAL4 to activate transcription, this effect does not occur in the presence of glucose, resulting in the release of GAL4 from DNA

independently modulated, could clearly not be achieved by stimulating the *de novo* synthesis of the factor, which would simply result in more of it being present. Hence, in addition to its rapidity, the activation of pre-existing factor has the advantage of flexibility in potentially being able to generate different forms of the factor with different activities. It should be noted, however, that this effect can also be achieved, by alternative splicing of the RNA encoding the factor (Section 10.2.3), which can, for example, generate forms of the protein with and without the DNA-binding domain, as in the case of the Era-1 factor, with and without the activation domain, as in the case of CREM or Oct-2, or with and without the ligand-binding domain, as in the case of the thyroid hormone receptor.

10.4 CONCLUSIONS

In this chapter we have discussed how the regulation of gene expression by transcription factors is achieved both by the regulated synthesis and by the regulated activity of these factors. Although there are exceptions, the regulation of synthesis of a particular factor is used primarily in cases of factors that mediate tissue-specific or developmentally regulated gene expression where a factor is required in only a small proportion of cell types and is never required in most cell types. In contrast, however, the rapid induction of transcription in response to inducers of gene expression is achieved primarily by the activation of pre-existing inactive forms of transcription factors that are present in most cell types, since this process, although more metabolically expensive, provides the required rapidity in response.

Although these two processes have been discussed separately, it should not be thought that a given factor can be regulated only at the level of synthesis or at the level of activity. In fact, in many cases of inducible gene expression which involve activation of pre-existing factors, such activation is supplemented by the slower process of synthesizing new factor in response to the inducing agent. Thus, in the case of the stimulation of genes containing AP1 sites by phorbol esters discussed above (Section 10.3.4), the phorbol ester-induced increase in the DNA binding of pre-existing Jun protein is supplemented by increased synthesis of both Fos and Jun following phorbol ester treatment, and such newly synthesized Fos and Jun will clearly eventually become a major part of the increased AP1 activity observed following phorbol ester treatment (see Section 7.2.1). Similarly, the activation of NFκB by dissociation from IκB following treatment with substances such as phorbol esters which activate T cells (see Sections 10.3.3 and 10.3.4) has been shown to be supplemented by increased synthesis of NFκB and its corresponding mRNA following T cell activation. Hence, in these cases, the rapid effects of post-translational processes in activating gene expression are supplemented by *de novo* synthesis of the factor which, although slower, will enhance and maintain the effect.

This combination of regulated synthesis and regulated activity is also seen in the case of factors that mediate tissue-specific gene expression and which are synthesized in only a few cell types. Thus, in the case of the MyoD factor which regulates muscle-specific genes, the factor and its corresponding mRNA are synthesized only in cells of the muscle lineage (see Section 5.3.2). The activation of MyoD-dependent genes that occurs when myoblast cells within this lineage differentiate into myotubes is not, however, mediated by new synthesis of MyoD, which is present at equal levels in both cell types. Rather, it occurs due to the decline in the level of the inhibitory protein Id, resulting in the post-transcriptional activation of pre-existing MyoD and the transcription of MyoD-dependent genes. Hence, in this case, regulation of synthesis is used to avoid the wasteful production of MyoD in cells of non-muscle lineage, whilst the activation of pre-existing MyoD ensures a rapid response to agents that induce differentiation within cells of the muscle lineage. Thus, in a number of cases, a combination of both regulated synthesis and regulated activity allows the precise requirements of a particular response to be fulfilled rapidly but with minimum unnecessary wastage of energy.

In summary, therefore, the different properties of regulated synthesis and regulated activity allow these two processes, both independently and in combination, to regulate efficiently the complex processes of inducible, tissue-specific and developmentally regulated gene expression.

REFERENCES

Altmann, M. and Trachsel, H. (1993). Regulation of translational initiation and modulation of cellular physiology. *Trends in Biochemical Sciences* **18**, 429–432.

Baeurele, P. A. (1995). Enter a polypeptide messenger. *Nature* **373**, 661–662.

Baniahmad, A., Tsai, S. Y., O'Malley, B. W. and Tsai, M.-J. (1992). Kindred S thyroid hormone receptor is an active and constitutive silencer and a repressor for thyroid hormone and retinoic acid responses. *Proceedings of the National Academy of Sciences USA* **89**, 10633–10637.

Beg, A. and Baldwin, A. S. (1993). The IκB proteins: multi functional regulators of Rel/NFκB transcription factors. *Genes and Development* **7**, 2064–2070.

Blank, V., Kowitsky, P. and Israel. A (1992). NFκB and related proteins: Rel/dorsal homologies meet ankyrin-like repeats. *Trends in Biochemical Sciences* **17**, 135–140.

Boyle, W. J., Smeal, T., Defize, L. H. K., Angel, P., Woodgett, J. R., Karin, M. and Hunter, T. (1991). Activation of protein kinase C decreases phosphorylation of c Jun at sites that negatively regulate its DNA binding activity. *Cell* **64**, 573–584.

Bresnick, E. H., Dalman, F. C., Sanchez, E. R. and Pratt, W. B. (1989). Evidence that the 90kDa heat shock protein is necessary for the steroid binding conformation of the L cell glucocorticoid receptor. *Journal of Biological Chemistry* **264**, 4992–4997.

Brown, K., Gerstberger, S., Carlson, L., Franzoso, G. and Siebenlist, U. (1995). Control of IκBα proteolysis by site-specific, signal-induced phosphorylation. *Science* **267**, 1485–1488.

Darnell, J. E. Jr, Kerr, I. M. and Stark, G. R. (1994). Jak–STAT pathways and transcriptional activation in response to IFNs and other extracellular signalling proteins. *Science* **264**, 1415–1421.

Davis, H. L., Weintraub, H. and Lassar, A. B. (1987). Expression of a single transfected cDNA converts fibroblasts to myoblasts. *Cell* **51**, 987–1000.

Descombes, P. and Schibler, U. (1991). A liver transcriptional activator protein LAP and a transcriptional inhibitory protein LIP are translated from the same mRNA. *Cell* **67**, 569–580.

Driever, W. and Nusslein-Volhard, C. (1988). The bicoid protein determines position in the *Drosophila* embryo in a concentration-dependent manner. *Cell* **54**, 95–104.

Gille, H., Sharrocks, A. D. and Shaw, P. E. (1992). Phosphorylation of transcription factor p62^{TCF} by MAP kinase stimulates ternary complex formation at c-*fos* promoter. *Nature* **358**, 414–417.

Gilmore, T. D., and Morin, P. J. (1993). The IκB proteins: members of a multi-functional family. *Trends in Genetics* **9**, 427–433.

Giniger, E., Varnam, S. M. and Ptashne, M. (1985). Specific DNA binding of GAL4, a positive regulatory protein of yeast. *Cell* **40**, 767–774.

Grimm, S. and Baeurele, P. A. (1993). The inducible transcription factor NFκB: structure–function relationship of its protein subunits. *Biochemical Journal* **290**, 297–308.

He, J. and Furmanski, P. (1995). Sequence specificity and transcriptional activation in the binding of lactoferrin to DNA. *Nature* **373**, 721–724.

Hill, C. S. and Treisman, R. (1995). Transcriptional regulation by extracellular signals: mechanisms and specificity. *Cell* **80**, 199–211.

Hunter, T. and Karin, M. (1992). The regulation of transcription by phosphorylation. *Cell* **70**, 375–387.

Ihle, J. N. and Kerr, I. M. (1995). Jaks and Stats in signalling by the cytokine receptor super-family. *Trends in Genetics* **11**, 69–74.

Ingham, P. W. (1988). The molecular genetics of embryonic pattern formation in *Drosophila*. *Nature* **335**, 25–34.

Jackson, S. P. and Tjian, R. (1988). O-glycosylation of eukaryotic transcription factors: implications for mechanisms of transcriptional regulation. *Cell* **55**, 125–133.

Jones, N. (1990). Transcriptional regulation by dimerization: two sides to an incestuous relationsip. *Cell* **61**, 9–11.

Karin, M. (1992). Signal transduction from cell surface to nucleus in development and disease. *FASEB Journal* **6**, 2581–2590.

Karin, M. and Smeal, T.(1992). Control of transcription factors by signal transduction pathways the beginning of the end. *Trends in Biochemical Sciences* **12**, 464–469.

Lalli, E. and Sassone-Corsi, P. (1994). Signal transduction and gene regulation: the nuclear response to cyclic AMP. *Journal of Biological Chemistry* **269**, 17359–17362.

Lamb, P. and McKnight, S. L. (1991). Diversity and specificity in transcriptional regulation: the benefits of heterotypic dimerization. *Trends in Biochemical Sciences* **16**, 417–422.

Larosa, G. J. and Gudas, L. J. (1988). Early retinoic acid-induced F9 teratocarcinoma stem cell gene ERA-1: alternative splicing creates transcripts for a homeobox-containing protein and one lacking the homeobox. *Molecular and Cellular Biology* **8**, 3906–3917.

Latchman, D. S. (1990). Cell-type specific splicing factors and the regulation of alternative RNA splicing. *New Biologist* **2**, 297–303.

Latchman, D. S. (1995). *Gene Regulation: A Eukaryotic Perspective*, 2nd edn. London: Chapman and Hall.

Lehming, N., Thanos, D., Brickman, J. M., Ma, J., Maniatis, T. and Ptashne, M. (1994). An HMG-like protein that can switch a transcriptional activator to a repressor. *Nature* **371**, 175–179.

Lillycrop, K. A. and Latchman, D. S. (1992). Alternative splicing of the Oct-2 transcription factor is differentially regulated in neuronal cells and B cells and results in protein isoforms with opposite effects on the activity of octamer/TAATGARAT-containing promoters. *Journal of Biological Chemistry* **267**, 24960–24966.

Lillycrop, K. A., Dawson, S. J., Estridge, J. K., Gerster, T., Matthias, P. and Latchman, D. S. (1994). Repression of a herpes simplex virus immediate-early promoter by the Oct-2 transcription factor is dependent upon an inhibitory region at the N-terminus of the protein. *Molecular and Cellular Biology* **14**, 7633–7642.

Luscher, B., Christenson, E., Litchfield, D. W., Knebs, E. G. and Eiserman, R. N. (1990). *Myb* DNA binding inhibited by phosphorylation at a site deleted during oncogenic activation. *Nature* **344**, 517–522.

McKeown, M. (1992). Alternative mRNA splicing. *Annual Review of Cell Biology* **8**, 133–155.

Manak, J. R., de Bisschop, N., Kris, R. M. and Prywes, R. (1990). Casein kinase II enhances the DNA binding ability of serum response factor. *Genes and Development* **4**, 955–967.

Nigg, E. A., Hilz, H., Eppenberger, H. M. and Dutly, F. (1985). Rapid and reversible translocation of the catalytic subunit of cAMP dependent protein kinase type II from the Golgi complex to the nucleus. *EMBO Journal* **4**, 2801–2806.

Palombella, V. J., Rando, O. J., Goldberg, A. L. and Maniatis, T. (1994). The ubiquitin-proteolytic pathway is required for processing the NF-κB1 precursor protein and the activation of NFκB. *Cell* **78**, 733–785.

Pratt, W. B. (1993). The role of heat shock proteins in regulating the function, folding and trafficking of the glucocorticoid receptor. *Journal of Biological Chemistry* **269**, 21455–21458.

Ronchi, E., Treisman, J., Dostatri, N., Struhl, G. and Desplan, C. (1993). Down-regulation of the *Drosophila* morphogen bicoid by the torso receptor mediated signal transduction cascade. *Cell* **74**, 347–355.

Sassone-Corsi, P. (1994). Goals for signal transduction pathways: linking up with transcriptional regulation. *EMBO Journal* **13**, 4628–4717.

Scheidereit, C., Heguy, A. and Roeder, R. G. (1987). Identification and purification of a human lymphoid-specific octamer-binding protein (OTF–2) that activates transcription of an immunoglobulin promoter *in vitro*. *Cell* **51**, 783–793.

Shen, W.-F., Detmer, K., Simonitch-Easton, T., Lawrence, H. J. and Largman, C. (1991). Alternative splicing of the Hox 2.2 homeobox gene in human hematopoetic cells and murine embryonic and adult tissues. *Nucleic Acids Research* **19**, 539–545.

Smith, D. F. and Toft, D. O. (1993). Steroid receptors and their associated proteins. *Molecular Endocrinology* **7**, 8–11.

Staudt, L. M., Clerc, R. G., Singh, H., Le Bowitz, J. H., Sharp, P. A. and Baltimore, D. (1988). Cloning of a lymphoid-specific cDNA encoding a protein binding the regulatory octamer DNA motif. *Science* **241**, 577–580.

Stevens CF (1994). CREB and memory consolidation. *Neuron* **13**, 769–770.

Thanos, D. and Maniatis, T. (1995). NFκB: a lesson in family values. *Cell* **80**, 529–532.

Thiele, D. J. (1992). Metal regulated transcription in eukaryotes. *Nucleic Acids Research* **20**, 1183–1191.

Treacy, M. N., Neilson, L. I., Turner, E. E., He, X. and Rosenfeld, M. G. (1992). Twin of I-POU: a two amino acid difference distinguishes an activator from an inhibitor of transcription. *Cell* **68**, 491–505.

Wuarin, J. and Schibler, U. (1990). Expression of the liver-enriched transcriptional activator protein DBP follows a stringent circadian rhythm. *Cell* **63**, 1257–1269.

Xanthopoulos, K. G., Mirkovitch, J., Decker, T., Kuo, C. G. and Darnell, J. E. Jr. (1989). Cell-specific transcriptional control of the mouse DNA binding protein mC/EBP. *Proceedings of the National Academy of Sciences USA* **86**, 4117–4121.

Yamamoto, K. R., Gonzalez, G. A., Menzel, P., Rivier, J. and Montminy, M. R. (1990). Characterization of a bipartite activation domain in transcription factor CREB. *Cell* **60**, 611–617.

Zimarino, V. and Wu, C. (1987). Induction of sequence-specific binding of *Drosophila* heat-shock activator protein without protein synthesis. *Nature* **327**, 727–730.

CHAPTER ELEVEN

Conclusions and future prospects

At the time the first edition of this book was published (1991) enormous progress had been made in understanding the nature and role of transcription factors. Thus the roles of specific factors in processes such as constitutive (Chapter 3), inducible (Chapter 4), tissue-specific (Chapter 5) and developmentally regulated (Chapter 6) gene expression had been defined, as had their involvement in diseases such as cancer (Chapter 7). Moreover, by studying these factors in detail, it proved possible to analyse how they fulfil their function in these processes by binding to specific sites in the DNA of regulated genes (Chapter 8) and activating or repressing transcription (Chapter 9), as well as the regulatory processes that result in their doing so only at the appropriate time and place (Chapter 10). Moreover, the regions of individual factors which mediate these effects and the critical amino acids within them that are of importance had been identified in a number of cases.

In the intervening years, much further progress has been made in these areas. In addition, the ability to prepare 'knock-out' mice, in which the gene encoding an individual factor has been inactivated, has allowed the *in vivo* functional role of many factors to be assessed directly, whilst numerous studies have elucidated the structure of specific factors, either in isolation or bound to DNA. It has become increasingly clear, however, that the activity of a particular factor cannot be considered in isolation. Thus, not only does the ability of a factor to stimulate transcription involve its ability to interact with components of the basal transcriptional complex (Section 9.2), but its ability to do this will depend on the presence of other factors in the cell. In part, the recognition of this fact has resulted from an increasing number of cases of factors that act to inhibit gene expression either directly or indirectly by interfering with the activity of a positively acting factor (see Section 9.3).

Additionally, however, it has become clear that, as well as stimulating or inhibiting factor activity, such protein–protein interactions can also alter the specificity of a factor. This may involve altering its DNA-binding specificity and hence the target genes it affects, as in the case of the interactions between the yeast a1 or MCMI proteins and the α2 protein (see Section 5.4.3). Alternatively, it may completely change the factor from activator to repressor, as in the case of the dorsal–DSP1 interaction (see Section 10.3.3). Indeed, differences in the ability to interact with other proteins result in factors with identical DNA-binding specificities having entirely different functional effects, as in the case of Ubx and Antp (see Section 6.2.3). Hence, by altering the specificity of particular factors, interactions of this type are likely to play a crucial role in the complex regulatory networks that allow a relatively small number of transcription factors to control highly complex processes such as development (see Chapter 6).

Ultimately, therefore, the understanding of transcription factor function will require a knowledge of the nature and effect of interactions between different factors which is as good as that now available for individual factors. Clearly, much work remains to be done before this is achieved. The rapid progress since the first edition of this book was published suggests, however, that an eventual understanding in molecular terms of the manner in which transcription factors control highly complex processes such as *Drosophila* and even mammalian development can ultimately be achieved.

Index

a and α genes *see* yeast mating types
A residues, interference analysis 28
Abd-B 154
Acanthamoeba, transcription 47–8, 50, 52
ACE1 factor, activation by copper 295–6
acidic domains 242–3
activation of transcription
    activation domains 239–56
      and chromatin structure 259
      functional relationships of domains
        244–6
      GAL4 acidic domain 244
      structures 242
    domain swapping 240
    inhibition
      by formation of non-DNA binding
        complex 266–7
      by masking of DNA binding site 264–6
    interaction with basal transcriptional
      complex 246–56
    multistage activation of transcription
      263–4
    other targets 257–62
adaptor molecules, interaction with
    activator molecules 256
adipocytes, differentiation 63
alcohol dehydrogenase (adh) gene 140
antennapedia gene *(antp)* 135–41, 143–5,
    152
    antp–btx complex 153
    homeodomain structure 202
    inhibition by Ubx 143–5

promoter 139–40
antennapedia protein (Antp) 121, 202
anti-oncogenes 185–92
    p53 185, 186–8, 228
    retinoblastoma gene *(rb-1)* 189–92
    Wilms' tumour gene product 185, 186
AP1 (activator protein-1) 72, 167–72
    DNA binding 267
    protein sites 305
      Fos protein 226–7
AP2
    binding sites 7
    DNA binding 228
assays
    DNA mobility shift assay 17–20
    DNAse I footprinting assay 28–9
    gel retardation assay 18
    methylation interference assay 25–8
    methylation protection assay 28–9
    in vivo footprinting assay 28–9
ATF, activation of transcription 247
ATGCAAAT, octamer motif 7, 8, 19–20
avian erythroblastosis virus (AEV) 172
    see also erbA
avian sarcoma virus (ASV) 167
5-azacytidine 107–8

B cell-specific enhancer 8
B cells
    expression of immunoglobulin genes
      101–2
      promoter and enhancer elements 102–7